Strengthening European Mobility Policy

"This book plays a crucial role in supporting the transformation towards a more sustainable mobility and logistics system through its interdisciplinary approach and by formulating concrete policy recommendations. It is written for a broader audience making it accessible to policymakers and practitioners. By bridging the polarity between technological and human perspectives, this volume paves the way for real change."
—Cathy Macharis, *Professor, Coordinator of the House of Sustainable Transitions, Vrije Universiteit Brussel, Belgium*

"Read this book to understand how engineers and social scientists can and should work together to support sustainable mobility transitions. The book offers fresh perspectives and innovative approaches for strategic planning and participatory planning, crucial for the successful implementation of inclusive mobility policies, low-carbon technologies, and new mobility services."
—Karst T. Geurs, *Professor, University of Twente, The Netherlands*

Imre Keseru · Samyajit Basu ·
Marianne Ryghaug · Tomas Moe Skjølsvold
Editors

Strengthening European Mobility Policy

Governance Recommendations from Innovative
Interdisciplinary Collaborations

Editors
Imre Keseru
Mobilise Mobility and Logistics Research Group, House of Sustainable Transitions (HOST)
Vrije Universiteit Brussel
Brussels, Belgium

Samyajit Basu
Mobilise Mobility and Logistics Research Group, House of Sustainable Transitions (HOST)
Vrije Universiteit Brussel
Brussels, Belgium

Marianne Ryghaug
Department of Interdisciplinary Studies of Culture
Norwegian University of Science and Technology
Trondheim, Norway

Tomas Moe Skjølsvold
Department of Interdisciplinary Studies of Culture
Norwegian University of Science and Technology
Trondheim, Norway

ISBN 978-3-031-67935-3   ISBN 978-3-031-67936-0 (eBook)
https://doi.org/10.1007/978-3-031-67936-0

© The Editor(s) (if applicable) and The Author(s) 2024. This book is an open access publication.

Open Access This book is licensed under the terms of the Creative Commons Attribution 4.0 International License (http://creativecommons.org/licenses/by/4.0/), which permits use, sharing, adaptation, distribution and reproduction in any medium or format, as long as you give appropriate credit to the original author(s) and the source, provide a link to the Creative Commons license and indicate if changes were made.
The images or other third party material in this book are included in the book's Creative Commons license, unless indicated otherwise in a credit line to the material. If material is not included in the book's Creative Commons license and your intended use is not permitted by statutory regulation or exceeds the permitted use, you will need to obtain permission directly from the copyright holder.
The use of general descriptive names, registered names, trademarks, service marks, etc. in this publication does not imply, even in the absence of a specific statement, that such names are exempt from the relevant protective laws and regulations and therefore free for general use.
The publisher, the authors and the editors are safe to assume that the advice and information in this book are believed to be true and accurate at the date of publication. Neither the publisher nor the authors or the editors give a warranty, expressed or implied, with respect to the material contained herein or for any errors or omissions that may have been made. The publisher remains neutral with regard to jurisdictional claims in published maps and institutional affiliations.

Cover illustration: © Melisa Hasan

This Palgrave Macmillan imprint is published by the registered company Springer Nature Switzerland AG
The registered company address is: Gewerbestrasse 11, 6330 Cham, Switzerland

If disposing of this product, please recycle the paper.

Foreword by Rosalinde van der Viles

Low-Carbon Approaches at the Crossroads: Why the European Green Deal Will Benefit from Interdisciplinary Insights

The European Union has outlined its ambitions to become the first climate neutral continent. The achievement of this ambition is supported through the EU Green Deal which sets out a long-term roadmap to deliver on the long-term systemic changes required. The roadmap covers a range of activities across sectors, including climate, energy, and mobility. At the heart of the EU Green Deal is the commitment to put people first and leave no person (or region) behind.

The contribution of Social Sciences and Humanities (SSH) research to low-carbon transitions cannot be understated. I am firmly convinced that we will fail in delivering upon our climate neutrality ambitions without SSH. SSH contribute to low-carbon transitions in multiple ways, including the development of inclusive approaches, the establishment of effective communication, and the creation of appropriate governance structures.

SSH research supports the establishment of an inclusive approach to achieving the EU's climate neutrality ambitions. As mentioned previously, a key component of the EU's approach to achieving climate neutrality is that no-one, and no region, is left behind. Inclusiveness is therefore important to ensure potential disparities and inequalities are addressed.

SSH support the establishment of inclusive practices by providing insight into cultural factors, such as values, beliefs, and identities and how these can support green policies and green transitions. Understanding the different cultures and experiences of individuals is a key component of ensuring no-one is left behind.

SSH also support communication and inform the development of effective public engagement initiatives. The incorporation of SSH insights into the narratives of transition can help convey to the public how low-carbon transitions are beneficial for the planet, the health and well-being of individuals, and the economy. SSH also help to demonstrate the necessity of behaviour changes to deliver on climate change. It is vitally important that policymakers get support on how to convey messages of urgency—but also the benefits on the lives of individual people—related to low-carbon policies. Not only does SSH research support effective communication related to low-carbon behaviours and practices, but it also provides insights on how changes will both be experienced and encouraged.

SSH research does not focus solely on behaviours and societal configurations; it also provides insights related to policy structures, institutions, and industry. These understandings can inform practices and help ensure that the actions undertaken are as effective as possible.

While SSH research and insights play a vital role in achieving the EU's low-carbon ambitions, they will have the most impact when integrated with other disciplinary perspectives, including those from Science, Technology, Engineering, and Mathematics (STEM). Most solutions for achieving low-carbon ambitions are situated at a crossroads: these solutions are not linked to a single sector or disciplinary background, rather they overlap between sectors and require the integration of different knowledge and perspectives. Bringing together different experts and experiences when developing approaches is essential to find innovative approaches to tackle climate change, undertake the energy transition, and establish sustainable mobility. Yet, in order to achieve all this, there is the need to continue breaking the silos in which research is undertaken and communicated. As such, interdisciplinarity between SSH and STEM needs to be promoted and supported.

Not only is there the need for interdisciplinarity between research disciplines, but there is also the need to have collaboration and communication between different actors. Achievement of low-carbon ambitions requires interactions between SSH, policymakers, and more technical and naturalistic disciplines. Policymaking needs to become more comprehensive and interdisciplinary in order to advance transitions, avoid duplications, and maximise impact through involving different people. The interactions between policy and research are critical as research and innovation activities need to be supported by the regulatory framework, with the regulatory framework also needing to be aware of the research and innovation activities undertaken to enable updates. Within policymaking, we need to continue to break those silos, adopt more interdisciplinary approaches, and make sure to bring along the societal dimension of the transition.

The collective intelligence across the three SSH CENTRE books—bringing together more than 150 researchers from more than 23 countries—is inspiring. The collaborations underpinning the chapters show how we should be working and are a starting point for breaking down silos. I really believe that collaborations between Social Scientists, Humanities researchers, and researchers from more technical disciplines, are key to advance low-carbon transitions. In order to achieve climate neutrality, there are many challenges to overcome, but the insights presented within these chapters and the expertise of the chapter authors can support the establishment of effective solutions, help break down barriers, and accelerate pathways to a sustainable and prosperous future.

Brussels, Belgium   Rosalinde van der Vlies
                                              Clean Planet Directorate,
                                               European Commission

Rosalinde van der Vlies is the Director of the Clean Planet Directorate in the European Commission's Directorate-General for Research and Innovation, and Deputy Mission Manager of the Climate-Neutral and Smart Cities Mission. Before her appointment as Director, Rosalinde van der Vlies was the Head of Coordination & Interinstitutional Relations Unit, and acting Head of Communication & Citizens Unit. Previously she held positions in Directorate-General Environment, Directorate-General Justice and Home Affairs, and in the private office of Janez Potočnik, the European Commissioner for the environment. Before joining the European Commission, she worked as a competition lawyer in an international law firm in Brussels and was a part-time teacher at the Catholic University in Brussels.

Foreword by Siddharth Sareen

A Wild Ride: Transdisciplinary Research to Inform Policies for Sustainable Transport Transitions

The twenty-first century is an exciting time for the transport sector. With digitalisation, electrification, and increased recognition of the necessity of collective, active, and integrated transport modes, a renaissance is underway. The forced automobility dependence built into urban, suburban, and rural transport infrastructures alike over the past "century of the automobile" is finally being challenged with some conviction and sense of hope. The urgent challenge of climate change mitigation has put new winds in the sails of ideas and praxis for innovative low-energy solutions to the familiar problem of how people move.

Transport challenges have moved from fundamentally engineering problems (efficiency, speed, acceleration) and design issues (comfort, ergonomics, compactness) to governance questions. How can we move smoothly and sufficiently using shared, integrated transport infrastructures and modes with low-energy demand that can be powered by low-carbon sources, typically using clean electricity, not fossil fuels? How can transport practices shift, without disruption, to significantly contrasting configurations, given enduring car-centric legacies in built environments?

This book makes three bold assumptions. First, that these solutions come from working across the social sciences, humanities, and the more technical fields that have traditionally dealt with transport.

Second, that there is appetite among policymakers for evidence-based, critically informed reflections. And third, that there is value in novel transdisciplinary collaborations that enable new thinking and action. In this Foreword, I examine these desirable points of departure in turn, to articulate why such an endeavour must succeed, and how it can.

Interdisciplinarity in transport research: under urgency (which we face globally, with unequal but increasingly uncertain distribution as climate change impacts start spiralling), core insights from disciplines must deliver value by being cross-fertilised between silos. A sociologist or geographer, and a transport modeller or analyst, have different conceptions of time and space, and deliver value by drawing on diverse data and analytical protocols. Yet their enterprise is truly successful if their analyses are richly informed by each other's insights, if these insights push them to problematise their own assumptions, and if the results and discussions that ensue bring their work into closer, more meaningful entanglement. Like operators of a bus fleet and a shared bicycle scheme developing mobility hubs to enable inter-modal connection, transport researchers across disciplines are most effective when leveraging their work through interdisciplinary hubs. This applies both for thematic impact, and for continued development and renewal within each discipline, through iterative reflection aided by adjacent alterity.

Evidence-based, critically informed transport policymaking: there is a joke about evidence-based policy versus "policy-based evidence". Like all good jokes, it works because it contains a kernel of truth. We live in a big data era, where the noise of information overload renders control over quality and mainstream narratives notoriously susceptible to co-optation by incumbents with purchasing power. Transport policymakers are hardly immune to these biases, being targeted by influential lobbies (look at skyrocketing SUV sales!) and struggling to overcome the impoverishment of collective imagination by decades of automobility-oriented development (hence the ecomodern enthusiasm for massively lithium-intensive personal electric cars). Good ideas from scholars, that have to compete for attention with visions backed by cash-rich incumbents, need more than "truth" or "scientific objectivity" on their side. They must be put into play by creating terms of engagement through vehicles like this book and the emergent networks of knowledge and academic practice that enable such science advice.

FOREWORD BY SIDDHARTH SAREEN   xi

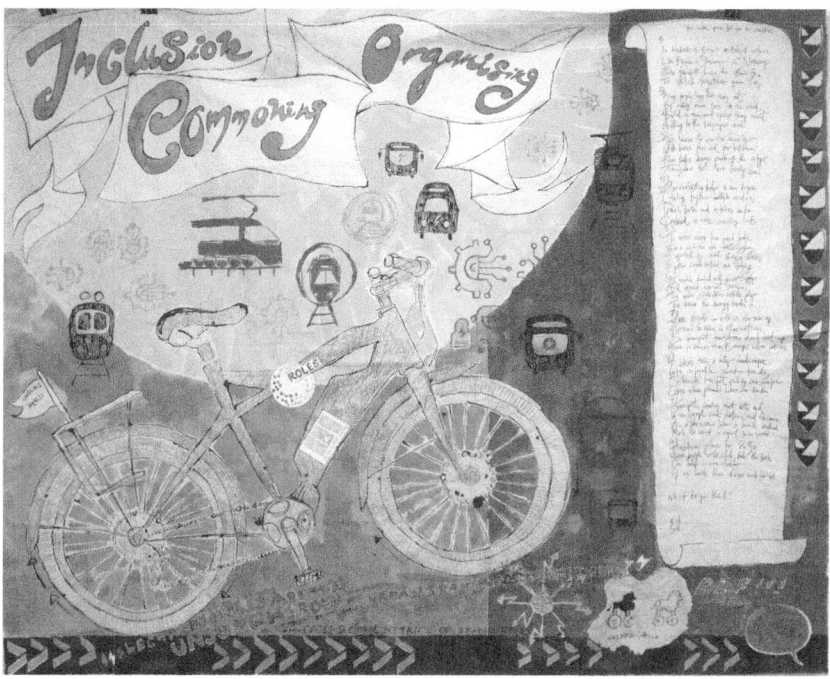

Fig.1 Electric bicycles are the unsung heroes of urban transport transitions. Source: Margrethe Brekke

Transdisciplinary collaborations: finally, I come to where the churn has to happen, for new thinking and action to enable and embody transitions to more hopeful, desirable transport futures. Figure 1 stems from my art-science collaboration with textile artist Margrethe Brekke, based on insights from the ROLES project on Responsive Organising for Low Emission Societies crystallised into three keywords: *inclusion, organising, and commoning*. It features my poem:

Responsive Organising

In hundreds of Europe's midsized cities
Like Bergen or Stavanger in Norway
Public transport lacks the density
To stitch together your day.

Many people buy their way out
By adding more cars to the road,
"Build us more road space," they shout,
Adding to the taxpayer's load.

This leaves the car-free minority
With buses few and far between,
Poor links across parts of the city,
Transfers that are rarely seen.

Digitalisation helps to some degree
Cobbling together multiple modes,
Trains, buses and e-bikes can be
Combined to ease commuting loads.

To move away from fossil fuels
These vehicles are electrifying,
Supported by smart charging tools,
System coordinators are trying

To match demand with power supply
And expand transport services,
The more collective vehicles ply
The lower the energy burden is.

Less private car batteries also ease up
Pressure to mine in other nations,
So transport transitions don't eat up
More resources than Europe's urban rations.

E-bikes turn a hilly landscape
Into enjoyable commuter tracks,
Sustainable transport policy can shape
Cities where planners have our backs.

Transport services must not end
When people seek nature and leisure,
An afternoon hike or beach weekend
Must be served in equal measure.

Ambitious plans for 2030
Where people walk, cycle, take the bus,
Can only become reality
If we make these choices work for us.

This book project is both an outcome of and a source of inspiration for, such joined-up thinking and action. May it mobilise us towards better ways of moving!

Oslo, Norway   Siddharth Sareen

Siddharth Sareen is a Professor in Energy and Environment at the Department of Media and Social Sciences, University of Stavanger, and Professor II at the Centre for Climate and Energy Transformation, University of Bergen. From August 2024 onwards, he is a Research Professor at the Fridtjof Nansen Institute in Oslo. His work focuses on the governance of energy transitions at multiple scales, in diverse contexts, and within and across a range of sectors, such as resource extraction, electricity generation, distribution and end-use, and urban transport. He is a board member of the Young Academy of Norway and the Empowered Futures Research School.

Preface

Research in the domains of transport and mobility inherently lies at the crossroads of multiple disciplines spanning the wide spectrum of Social Sciences and Humanities (SSH) and Science, Technology, Engineering, and Mathematics (STEM). Researchers in the fields of mobility and logistics often face a dilemma when submitting funding applications as transport rarely appears as a research category in application forms and they have to categorise their project proposals within other social or technical disciplines. Indeed, transport research borrows from and builds on methods and approaches from various other research fields. Still these methods and approaches ever too often remain confined within their own "silos" of technology and engineering or social and human sciences. The composition of the editorial team for this book reflects an effort to break these silos, bringing together editors from transport geography, transport engineering, political science, and sociology. We have endeavoured to create this book project with the aim to demonstrate that SSH-STEM collaborations can be successful in producing insights that are valuable both for research and policy.

This book is a key output from the Horizon Europe project SSH CENTRE: *Social Sciences and Humanities for Climate, Energy and Transport Research Excellence*. Each chapter represents findings from a novel collaboration between the technical and social sciences based on a call to form ad-hoc research teams and open for all topics related to the objectives of the Green Deal of the European Commission. This diversity is reflected in the wide range of themes covered in this book, yet all chapters are united by the common goals of demonstrating SSH-STEM collaboration and providing accessible policy recommendations for the European Commission grounded in research.

The research teams not only bridged the differences in working methods and terminologies between SSH and STEM research communities but also stepped out of the confines of academia to provide practical and actionable policy recommendations. This effort aims to address the often-present gap between research, policy, and practice. Fundamentally, this book is aimed at strengthening European mobility policy through better interdisciplinary research. It is part of a three-volume collection with the other volumes focusing on recommendations for climate policy and energy policy; all three volumes are available open access.

The Editors' time on this book—in addition to the collaboration expenses of the chapter teams—was funded by the SSH CENTRE project. This project is funded by the European Union's Horizon Europe research and innovation programme (under grant agreement no. 101069529) and by the UK Research and Innovation under the UK Government's Horizon Europe funding guarantee (grant no. 10038991).

This book would not have been possible without the tireless efforts of Ami Crowther at Anglia Ruskin University, who consistently ensured that the editor and author teams stayed on track and coordinated across the three sister volumes of this collection.

Brussels, Belgium	Imre Keseru
Brussels, Belgium	Samyajit Basu
Trondheim, Norway	Marianne Ryghaug
Trondheim, Norway	Tomas Moe Skjølsvold

Contents

Part I Introduction

1 Strengthening European Mobility Policy: Governance Recommendations from Innovative Interdisciplinary Collaborations 3
Imre Keseru, Samyajit Basu, Marianne Ryghaug, and Tomas Moe Skjølsvold

Part II The Role of Public Policy in Supporting Sustainable Mobility Transitions

2 Evaluating Public Policies for Sustainable Mobility: A Review Through Some Interdisciplinary Methodologies and Procedures 13
Francisco Alonso, Mireia Faus, Sergio A. Useche, José Luis Velarte, and Mónica Alonso

3 Enhancing Policy Coordination of Schooling and Transport for Net-Zero 27
Joshua Lait, Femke Nijsse, Stewart Barr, David Hall, Catherine Butler, Antonio Olmedo, and Cormac Lynch

4 Improving Social Justice, Environmental Integrity, and Geopolitical Resilience in EU Electric Mobility Transition 41
Aleksandra Lis-Plesińska, Nathalie Ortar, Rafał Szymanowski, Marek W. Jaskólski, Aleksandra Parteka, and Christine Buisson

Part III The Potential of the Transformation of Transport Services and Vehicle Technologies to Contribute to the Transition

5 Promoting Sustainable Urban Mobility Through Implementation of Electric Buses: A Case Study of Ostrava 59
Marek Krumnikl, Adam Červenka, Filip Lapuník, and Luboš Mikula

6 Improving Rural Quality of Life by Combining Public Transportation with Demand Responsive Transport Systems 73
József Pál Lieszkovszky, Dániel Tordai, Daniel Hörcher, Tamás Fleischer, and András Munkácsy

7 Providing State-Supported Financial Incentives and Benefits for Vehicle Insurance Policies Using Telematics 87
Virginia Petraki, Apostolos Ziakopoulos, Evangelia Fragkiadaki, Nikolaos Karouzakis, Konstantinos Kakavoulis, and George Yannis

Part IV New Methods and Tools to Support Participatory Planning

8 Assessing Mobility Policy with AI-Driven Analysis of User-Generated Content 103
Floriano Tori, Charlotte van Vessem, Juliana Betancur Arenas, and Vincent Ginis

9	Enabling Inclusive Urban Transport Planning Through Civic Artificial Intelligence Dimitris Michailidis, Kristina Khutsishvili, Konstantinos Konstantis, Aristotle Tympas, Imad Antoine Ibrahim, and Sennay Ghebreab	115
10	Facilitating Sustainable Logistics Policy Development Using Multicriteria Satisfaction Analysis: A Case of Preference Mapping for Cargo Bike Last-Mile Delivery He Huang, Xu Zhang, Salvatore Corrente, Sajid Siraj, and Maja Kiba-Janiak	129

Part V Conclusion

11	Recommendations for Future Interdisciplinary Collaborations Within Transport and Mobility Marianne Ryghaug, Tomas Moe Skjølsvold, Imre Keseru, and Samyajit Basu	147

Afterword 1: From Many Hands Problem to Unconscious Assumptions: Transforming Our Governance Systems	155
Afterword 2: What (About) Now? Complexities, Omissions, and Taking Transitions Seriously	159
Afterword 3: Cities in Transition	163
Afterword 4: Sustainable Mobility and Systemic Change: The Power of Collaborative Governance	167
Index	171

Notes on Contributors

Mónica Alonso is a civil engineer with a Ph.D. in Traffic and Road Safety from the Polytechnic University of Valencia, Spain.

Francisco Alonso is a full professor (profile "Traffic and Road Safety") and director of the University Research Institute on Traffic and Road Safety (INTRAS) at the University of Valencia, Spain. He has a Ph.D. in Decision-Making.

Juliana Betancur Arenas is a Ph.D. researcher at the Mobilise Mobility and Logistics Research Group and the House of Sustainable Transitions at the Vrije Universiteit Brussel, Belgium. Her research focuses on active mobility adoption processes, from infrastructure to bike users' behaviour.

Stewart Barr is professor of Geography at the University of Exeter. His research interests focus on personal mobility and behavioural change within the context of sustainable development and climate change.

Samyajit Basu is a senior researcher at the Mobilise Mobility and Logistics Research Group and the House of Sustainable Transitions at the Vrije Universiteit Brussel, Belgium. His primary research interests are sustainable transport, accessible and inclusive mobility, transport policy, road safety, human factors and just transition in transportation.

Christine Buisson is a senior researcher at Gustave Eiffel University (LICIT-ECO7 laboratory). Her research focuses on road and rail flows.

She analyses detailed datasets and develops methodological critics of traffic models.

Catherine Butler is an associate professor of Human Geography at the University of Exeter. Her research centres around the analysis of environmental issues and societal responses to address them.

Mgr. Adam Červenka is a researcher specialising in urban development, focusing on shrinking cities, walkability, and improving urban spaces. He is experienced in spatial data analysis and modern Geographic Information System (GIS) tools.

Salvatore Corrente is a full professor at the Department of Economics and Business at the University of Catania. His research interests concern Multiple Criteria Decision-Making and Interactive Evolutionary Multiobjective Optimization.

Roberta Dall'Olio is the Director of European Association of Development Agencies (EURADA). Her assignments include the design and (technical) management of international and territorial cooperation on institution and capacity building, regional and local development, social innovation, culture and creativity, as well as urban regeneration. Roberta graduated in Law and European Union Policies. She is a Member of various EC working groups (in DG Regio and DG Grow). She is a member of the Business Advisory Board of the SSH-Centre project.

Mireia Faus is a researcher with teaching responsibilities, attached to the University Research Institute on Traffic and Road Safety (INTRAS) at the University of Valencia, Spain. Mireia has a Ph.D. in Psychology Research.

Tamás Fleischer is a civil engineer, economist, and former researcher at the Centre for Economic and Regional Studies of the Hungarian Academy of Sciences. His research focuses on infrastructure networks, especially environment-friendly transport solutions.

Evangelia Fragkiadaki is a senior lecturer and programme leader in Counselling Psychology at the University of the West of England, Bristol. Her research explores psychological interventions that meet the real needs of the community.

Vincent Ginis is an assistant professor of Mathematics at the Vrije Universiteit Brussel and a visiting professor at Harvard University. His research interests include data analytics, applied physics, and machine learning.

Sennay Ghebreab is group leader of the Socially Intelligent Artificial Systems (SIAS) group at the University of Amsterdam, and scientific director of Civic AI Lab at the National Centre for AI, the Netherlands.

Oliver Greenfield is the Convenor of the Green Economy Coalition, the world's largest multi-stakeholder alliance committed to accelerating the transition to green economies by building evidence and societal demand. Oliver has degrees in engineering, business, and economics. He is a member of the Business Advisory Board of the SSH-Centre project.

David Hall is professor of Education at the University of Exeter. His research focuses upon large-scale education reforms and their implications for education institutions and those that work in them.

Debbie Hopkins is an associate professor in Human Geography at the University of Oxford (UK). She works between the School of Geography and the Environment and the Sustainable Urban Development programme in the Department for Continuing Education. Debbie leads a UKRI Economic and Social Research Council project, "Trucking Lives", and has research interests across mobile working lives and their entanglements with decarbonisation.

Daniel Hörcher is a postdoctoral research associate at the Department of Civil and Environmental Engineering at Imperial College London. He is a transport scientist with a track record in public transport research.

He Huang is a scientist at the Paul Scherrer Institute (PSI) in Switzerland and the Mobilise Mobility and Logistics Research Group at the Vrije Universiteit Brussel, Belgium. Specialising in Multi-Criteria Decision-Making (MCDM), his research scope also includes group decision-making and optimisation.

Imad Antoine Ibrahim is an assistant professor in Environmental Law & Governance at the University of Twente, Faculty of Behavioural, Management and Social Sciences, Section of Governance and Technology for Sustainability.

Marek W. Jaskólski is a geographer-practitioner in urban planning, using approaches from the borderline of Earth sciences, socio-economic geography, and spatial economy, interested in practical applications of the idea of sustainable development.

Konstantinos Kakavoulis is a founding partner at Digital Law Experts (DLE), a niche law firm based in Athens, Greece. Being a pioneer in digital rights protection in Greece, he has also co-founded Homo Digitalis, the first digital rights organisation in the country.

Nikolaos Karouzakis is an assistant professor of finance at Alba Graduate Business School, Greece. He received his Ph.D. from Bayes Business School and worked as a postdoctoral researcher at the London School of Economics and Political Science.

Imre Keseru is an assistant professor of urban mobility and deputy director at the Mobilise Mobility and Logistics Research Group, part of the House of Sustainable Transitions at the Vrije Universitet Brussel, Belgium. His research focuses on participatory transport planning, foresight, and stakeholder-based evaluation of mobility.

Kristina Khutsishvili is a postdoctoral researcher in AI and Society based in Amsterdam. Her experiences include leading the work package 'Ethical and Inclusive Engagement in Practice' of the CommuniCity Horizon Europe project.

Maja Kiba-Janiak is an associate professor at Wroclaw University of Economics and Business. Her research interests concern the co-creation of sustainable city logistics, including the use of MCDM/A and qualitative methods.

Konstantinos Konstantis is a Ph.D. Candidate in AI Ethics at the Department of History and Philosophy of Science, School of Science, National and Kapodistrian University of Athens.

Mgr. Marek Krumnikl is a researcher specialising in sustainable mobility and policy. He is pursuing a Ph.D. in sustainable mobility for Eastern and Central European cities, focusing on urban planning and environmentally friendly transportation systems.

Joshua Lait is an environmental social scientist at the University of Exeter. His research interests focus on the impacts of sectoral and institutional policies on sustainable transitions.

Ing. Filip Lapuník is a researcher specialising in automotive engineering and transportation, with a focus on road vehicle aerodynamics. He is experienced in transport engineering design, materials, manufacturing, CAD software, and CFD.

József Pál Lieszkovszky is a geographer and social scientist, working as a senior researcher at the Transport Development Research Centre at KTI Hungarian Institute for Transport Sciences and Logistics.

Aleksandra Lis-Plesińska is an associate professor in Social Anthropology at the Adam Mickiewicz University in Poznań-Poland, working on socio-technical transitions of energy and transport systems, public perceptions of technological risks, science-policy nexus in climate-energy politics.

Cormac Lynch is a complexity economist using mixed methods approaches to research the social and macroeconomic effects of decarbonisation and economic change.

Dimitris Michailidis is a Ph.D. Candidate in Socially-Aware Artificial Intelligence at the University of Amsterdam, researching the role of AI in designing inclusive, accessible, and sustainable transport systems.

Ing. Luboš Mikula is a researcher specialising in autonomous vehicle safety and reliability, with a Masters in Transport and Functional Safety. He is a doctoral candidate at Banska Technical University, focusing on sustainable and efficient transportation systems.

Miloš N. Mladenović is associate professor at the Spatial Planning and Transportation Engineering group within a multidisciplinary Department of Built Environment at Aalto University (Finland). Miloš's research focuses on the transformation of urban mobility systems and their governance processes, in a close collaboration with practitioners across Europe, and drawing from a plethora of theories and methods across various STEM and SSH fields.

András Munkácsy is a senior researcher and head of the Transport Development Research Centre at KTI Hungarian Institute for Transport Sciences and Logistics in Budapest.

Femke Nijsse is a complexity scientist with an interest in quantitative modelling of energy systems, with a focus on the power sector, transport, and residential heating.

Antonio Olmedo is an associate professor in Education Policy Sociology. His research interests focus on neoliberal policies and the creation of quasi-markets in education: privatisation, competition, and school choice.

Nathalie Ortar is a senior researcher in anthropology at the ENTPE-University in Lyon, France. She is a specialist in urban and rural mobility issues and energy transition with a focus on energy and mobility justice.

Aleksandra Parteka is an associate professor at the Faculty of Management and Economics, Gdansk University of Technology, Poland, focusing on international economics, the impact of trade, integration, technological progress on labour markets and productivity.

Virginia Petraki is a Ph.D. Candidate at the Department of Transportation Planning and Engineering of the National Technical University of Athens, specialising in transport engineering, road safety, and transport economics.

Marianne Ryghaug is a professor of Science and Technology Studies at the Norwegian University of Science and Technology (NTNU) and senior researcher at SINTEF. She specialises in interdisciplinary research on socio-technical transitions in energy and transport, sustainable mobility, digitalization, justice, and public engagement.

Sajid Siraj is an associate professor at Leeds University Business School. He worked for several years in industry before pursuing an academic career in machine learning and decision support systems.

Tomas Moe Skjølsvold is a professor at the Norwegian University of Science and Technology (NTNU) and the director of the Norwegian Centre for Energy Transition Strategies (NTRANS). His key interests include the politics of transitions, the justice implications of transitions, and the social aspects of new technologies.

Rafał Szymanowski is an assistant professor at the Faculty of Political Science and Journalism at Adam Mickiewicz University in Poznań, Poland, working on political economy of economic crises, role of national economic ideas, and sustainable transport.

Dániel Tordai is a junior researcher at KTI Hungarian Institute for Transport Sciences and Logistics, where he focuses on transportation economics and policy. He is also pursuing a Ph.D. at the Budapest University of Technology and Economics.

Floriano Tori is a Ph.D. researcher at the Vrije Universiteit Brussel, Belgium. His research interests lie at the interface between Physics and Machine Learning.

Aristotle Tympas is a professor of the History of Technology and Director of the 'Science, Technology, Society—Science and Technology Studies' graduate programme at the National and Kapodistrian University of Athens.

Sergio A. Useche is a professor in the Department of Basic Psychology and a researcher at the University Research Institute on Traffic and Road Safety (INTRAS) at the University of Valencia, Spain. He has a Ph.D. in Psychology Research.

José Luis Velarte is a civil engineer with more than 10 years of experience as a university researcher, specialised in mobility and road safety at the University Research Institute on Traffic and Road Safety (INTRAS) at the University of Valencia, Spain.

Charlotte van Vessem is a Ph.D. researcher at the Mobilise Mobility and Logistics Research Group and the House of Sustainable Transitions at the Vrije Universiteit Brussel, Belgium. She researches gendered mobilities and 15-minute cities.

Prof. George Yannis is director of the Department of Transportation Planning Engineering at the National Technical University of Athens, with more than 30 years of expertise as engineer, academic, advisor, and decision-maker.

Dr. Apostolos Ziakopoulos is a research associate at the School of Civil Engineering of the National Technical University of Athens, with 9 years of experience in transport engineering and road safety.

Lucian Zagan is a project coordinator in the mobility team at Eurocities. Eurocities is the largest network of European cities, counting 200+ cities among its members. Lucian holds a master's degree in urban mobility from the University of Architecture and Urban Planning in Bucharest, and is currently pursuing a Ph.D. in planning and environmental policy at University College Dublin.

Xu Zhang is a Ph.D. researcher in the School of Transport and Civil Engineering at Technical University Dublin in Ireland. Her doctoral research focuses on sustainability assessment in urban logistics.

List of Figures

Fig. 5.1	Grams of CO_2/km produced based on energy consumption of the vehicle (Color figure online)	66
Fig. 8.1	Distribution of the labelled tweets by expert annotators (blue) and by GPT-4 (red) (color figure online)	108
Fig. 8.2	**I.** Distribution by date of creation of the total tweets collected. Peaks in the distribution can be correlated to known mobility interventions. The label indicates the difference in tweet types. **II.** Aggregated histograms of tweets labelled by GPT-4 for each labelling category (Mode mentioned, domain, and sentiment). **(III, IV, V, VI).** For each active and shared mobility mode we selected the tweets labelled by GPT-4 to contain that mode. For each of these subcategories of tweets, we show the distribution of the domain mentioned (in percentages of total tweets of the selected mode), and within the domains, we show sentiment percentages with pie charts	110
Fig. 9.1	A framework for inclusive transport planning, in which communities actively engage in the design and evaluation of the system used to generate transport projects	121
Fig. 10.1	MUSA overall relative action diagram	137
Fig. 10.2	MUSA overall relative improvement diagram	139

List of Tables

Table 4.1   Identified problems, recommendations to solve it,
             and the addressed audiences   50
Table 7.1   Social CBA for the implementation of telematics
             insurance policies in Greece   94
Table 10.1   Key criteria and survey questions   136

PART I

Introduction

CHAPTER 1

Strengthening European Mobility Policy: Governance Recommendations from Innovative Interdisciplinary Collaborations

Imre Keseru, Samyajit Basu, Marianne Ryghaug, and Tomas Moe Skjølsvold

Abstract This introduction chapter provides context and definitions to support engagement with the book, including current EU mobility ambitions, our understanding of interdisciplinarity, and the aims and purpose of the book project.

I. Keseru (✉) · S. Basu
House of Sustainable Transitions (HOST), Mobilise Mobility and Logistics Research Group, Vrije Universiteit Brussel, Brussels, Belgium
e-mail: imre.keseru@vub.be

S. Basu
e-mail: samyajit.basu@vub.be

M. Ryghaug · T. M. Skjølsvold
Department of Interdisciplinary Studies of Culture, Norwegian University of Science and Technology, Trondheim, Norway

© The Author(s) 2024
I. Keseru et al. (eds.), *Strengthening European Mobility Policy*,
https://doi.org/10.1007/978-3-031-67936-0_1

Keywords Interdisciplinarity · Socio-technical transitions · SSH-STEM collaboration · EU policy

CURRENT EU MOBILITY POLICY AMBITIONS

Since transport represents 25% of greenhouse gas emissions in the EU, one of the key targets of the European Green Deal is to cut carbon emissions in transport by 90% by 2050 (European Commission, 2019). The European Commission has also developed a "Sustainable and Smart Mobility Strategy—putting European transport on track for the future" (European Commission, 2020) which wants to improve the resilience of the transport system against future crises. The strategy also wants to make mobility available and affordable for all by making rural and remote regions better connected, and to make transport accessible for persons with reduced mobility and disabilities. In another ambitious initiative, 112 pioneer cities have been selected that committed to become climate neutral by 2030 (European Commission, 2021). Considering the significant contribution of transport to carbon emissions, especially in urban areas, decarbonising urban mobility will be at the heart of this challenge. The involvement of citizens as well as local businesses in this decarbonisation process is crucial to ensure that the low-carbon transition remains a "just transition" that can improve citizens' well-being equitably across the society by improving air quality, creating jobs, promoting healthier lifestyles, and reducing the negative effects of mobility.

This will entail social and technological changes that might disrupt dominant transport infrastructures and services, as well as having broad implications for future societies. On the one hand, current transport systems contain strong lock-ins of carbon-intensive industries and practices that perpetuate deep social, economic, and environmental challenges. On the other hand, transport systems may enable economic

---

e-mail: marianne.ryghaug@ntnu.no

T. M. Skjølsvold
e-mail: tomas.skjolsvold@ntnu.no

growth and provide access to essential services, education, employment, leisure, and health services. Mobility transitions, therefore, need to either challenge the problematic aspects of contemporary systems, while ensuring that overall levels of mobility services are maintained or improved, or seek to have an impact on the organisation of society to minimise the need for transport. Thus, there is a need to consider how these multiple objectives can be balanced while reducing existing inequalities and not creating new ones.

The Need for SSH-STEM Collaboration to Meet Complex European Mobility Challenges

The dominance of technical and engineering perspectives in transport planning and infrastructure design have been increasingly challenged over the past three decades (Ryghaug et al., 2023). A fundamental transformation has gained momentum from a "predict and provide" paradigm that aims to satisfy continuously increasing travel demand by building new or expanding existing infrastructure, towards an "avoid-shift-improve" approach (United Nations, 2016). This alternative paradigm represents a shift away from simply accommodating the growing demand for transportation infrastructure towards strategies that prioritise sustainability, efficiency, and quality of life. Therefore, there is an increasingly urgent need to investigate the human and social aspects of transport rather than focusing only on technology development, infrastructure design, and construction (Ryghaug et al., 2023).

The contributions of Social Sciences and Humanities (SSH) to transport and mobility research have significantly diversified and enriched the domain. These contributions have introduced substantial theoretical and empirical advancements, offering interdisciplinary knowledge essential for facilitating transitions to more sustainable mobility systems. Relevant SSH disciplines in this respect are Political Science (e.g. to better understand the complexities of decision-making, policy formulation, and governance within the transport sector), Human Geography (to account for spatial differences in needs, demand, and accessibility), Sociology (to explore factors that influence travel behaviour and account for social equity), Science and Technology Studies (to understand the reciprocal relationship and mutual shaping of technology and society), Anthropology (to discover how cultural factors affect travel choices), Economics (to account for the externalities of transport activities), Urban Planning (to integrate the development of urban spaces and transport infrastructure) among others. It is however not enough to "involve" SSH researchers in dominantly technological and technical projects. Rather, a meaningful collaboration across the spectrum of SSH and Science, Technology, Engineering, and Mathematics (STEM) disciplines will be needed for EU mobility policy to reach its goals.

Stimulating Novel SSH-STEM Collaborations Through This Book Project

This book focuses on developing policy recommendations that can support the transition towards a just, inclusive, smart, competitive, safe, accessible, green, and affordable transport system based on academic research conducted by research teams composed of SSH and STEM researchers. The book offers a new and unique perspective on some of the key challenges of the mobility transition by developing interdisciplinary approaches to propose policy recommendations to the European Commission.

This volume focuses on fostering and demonstrating SSH-STEM collaborative practices to develop EU-level policy, aligning with the objectives of the SSH CENTRE project. Each chapter has been researched and written by at least two SSH and two STEM researchers who collaborated in defining the research question, collecting and analysing data, interpreting the results, and phrasing policy recommendations. Consequently, the book, firstly serves to showcase and disseminate policy-relevant recommendations for mobility and related sectors to support the achievement of the EU's ambitions; and secondly, it created a kind of sandbox to experiment with the integration of SSH and STEM methods and knowledge demonstrated in the chapters of this book.

Overview of the Book Contents

The chapters in this book address three key aspects of the transition of the mobility system towards carbon neutrality. The first three chapters in Part II investigate the role of public policy in supporting sustainable mobility transitions, how the coordination of transport and other policies such as education can act as a transformative strategy, and how new combined qualitative-quantitative approaches can contribute to better policy evaluation. Chapter 2 (Alonso et al.) focuses on how evaluation processes could be enhanced to support reaching the targets of the Green Deal by using new technologies as well as a set of key performance indicators across different projects to enable benchmarking and comparison. Chapter 3 (Lait et al.) revolves around the nexus of transportation and schooling policies, highlighting that liberalisation in both domains have resulted in the need to coordinate both actors and policies across these domains to reduce transport demand. Chapter 4 (Lis-Plesińska et al.) focuses on the complexity of mobility transitions in terms of social justice implications, environmental integrity, and geopolitical resilience. Synthesising across research projects, the chapter highlights how a broad combination of disciplines is needed to address and formulate viable recommendations in this space.

The third part of the book (Chapters 5–7) investigates the potential of the transformation of transport services and vehicle technologies to contribute to the transition. The contributions study the technological and organisational challenges of the transition and compare specific technologies or services. Chapter 5 (Krumnikl et al.) discusses how Ostrava is trying to promote sustainable urban mobility through the transition to electric and CNG buses, aligning with EU and national policies. It demonstrates success in fleet modernisation, with potential economic and ecological benefits, especially in regions with cleaner electricity generation. However, strategic planning and subsidy are essential for the successful implementation and long-term cost-effectiveness of such measures. Chapter 6 (Lieszkovszky et al.) explores strategies to improve access to essential services and life activities in rural areas by introducing innovative demand responsive transport (DRT) services. The authors put forward recommendations to address the challenges of long-term funding, enhanced collaboration across different transport providers, and reduce the barriers of entry to the DRT market by small and medium-sized companies and community initiatives. Chapter 7 (Petraki et al.) demonstrates how state-supported financial incentives can promote safe and eco-friendly driving behaviour through telematics-based vehicle insurance policies. Using social cost–benefit analysis (CBA) on a case study in Greece, authors show significant reductions in road casualties and environmental benefits with positive socio-economic indicators. It also highlights the crucial role of collaboration across disciplines for effective policy design and implementation.

Stakeholder participation is also a key aspect of a just transition. The authors of the chapters in Part IV (Chapters 8–10) offer new insights into methods and tools to support participatory planning through artificial intelligence and mass-participation. Chapter 8 (Tori et al.) explores how to assess mobility policy with AI-driven analysis of user-generated content. The authors argue that applying this methodology through recent open-source AI developments such as ChatGPT allows decision-makers and their teams to rapidly generate and assess a great amount of relevant data which can facilitate the effectiveness and efficiency of policymakers in the decision-making processes in urban mobility planning. Chapter 9 (Michailidis et al.) explores the potential role of artificial intelligence to assist in complex urban transport network design decisions, and support the participation of local citizens in addressing accessibility needs. The authors propose a framework for urban transport design through civic artificial intelligence via technology that directly integrates community preferences and feedback into AI training. Finally, Chapter 10 (Huang et al.) explores how multicriteria satisfaction analysis (MUSA) may be used to foster public participation for developing policies for cargo bike deliveries in European cities. The MUSA Average Satisfaction Indices help to visualise perceptions of citizens towards sustainable last-mile delivery initiatives, which, according to the authors, may provide evidence-based support for local authorities and city managers to get a more nuanced view of their community and neighbourhood.

Tips on How to Read This Book

This book is best read as an experiment in interdisciplinarity with clear policy ambitions. Reading it from cover-to-cover might not prove the smoothest reading experience, but as snapshots into creative processes seeking to stretch current research and impact practices there is much to be gleaned from the chapters at hand for researchers in the field. Similarly, policymakers who are seeking fresh thoughts on how to tackle concrete challenges, or becoming sensitised to challenges they did not know they were up against, similarly might find strong inspiration.

The book's chapters are intentionally short and focus on an overarching policy recommendation with supporting evidence but without an extensive literature review (unless the main method involved a review) or methodology sections. At the beginning of each chapter, there are a series of 'policy highlights'. These summarise the chapter's policy recommendations and reference the interdisciplinary activities that informed their development. The conclusion section of each chapter then further elaborates on them. Some of the chapters make reference to appendices - these additional materials have been uploaded to the SSH CENTRE's Zenodo site. The conclusion chapter (Chapter 11) reflects on the policy recommendations and the process of identifying them through the interdisciplinary collaborations.

The forewords and afterwords framing this book promote dialogue on STEM-SSH (and broader) collaborations for low-carbon mobility futures in Europe. The forewords offer perspectives from policy emphasising the significance of SSH-STEM collaboration. The afterwords, contributed by SSH and STEM researchers, policy actors, and members of the SSH CENTRE project's business advisory board, reflect on the policy recommendations presented in the book, grounding their reflections in their own experiences and insights.

REFERENCES

European Commission. (2019). *Communication from the Commission to the European Parliament, the European Council, the Council, the European Economic and Social Committee and the Committee of the Regions.* The European Green Deal. COM/2019/640 final. https://eur-lex.europa.eu/legal-content/EN/TXT/?uri=COM%3A2019%3A640%3AFIN

European Commission. (2020). *Communication from the Commission to the European Parliament, the Council, the European Economic and Social Committee and the Committee of the Regions.* Sustainable and Smart Mobility Strategy—Putting European transport on track for the future. COM/2020/789 final. https://eur-lex.europa.eu/legal-content/EN/TXT/?uri=CELEX%3A52020DC0789

European Commission. (2021). *Communication from the Commission to the European Parliament, the Council, the European Economic and Social Committee and the Committee of the Regions on European Missions.* COM/2021/609 final. https://eur-lex.europa.eu/legal-content/EN/TXT/?uri=CELEX%3A52021DC0609

Ryghaug, M., Subotički, I., Smeds, E., von Wirth, T., Scherrer, A., Foulds, C., Wentland, A., et al. (2023). A Social Sciences and Humanities research agenda for transport and mobility in Europe: Key themes and 100 research questions. *Transport Reviews, 43*(4), 755–779.

United Nations. (2016). *Mobilizing sustainable transport for development: Analysis and policy recommendations from the United Nations Secretary-General's high level advisory group on sustainable transport. technology.* https://sustainabledevelopment.un.org/content/documents/2375Mobilizing%20Sustainable%20Transport.pdf

Open Access This chapter is licensed under the terms of the Creative Commons Attribution 4.0 International License (http://creativecommons.org/licenses/by/4.0/), which permits use, sharing, adaptation, distribution and reproduction in any medium or format, as long as you give appropriate credit to the original author(s) and the source, provide a link to the Creative Commons license and indicate if changes were made.

The images or other third party material in this chapter are included in the chapter's Creative Commons license, unless indicated otherwise in a credit line to the material. If material is not included in the chapter's Creative Commons license and your intended use is not permitted by statutory regulation or exceeds the permitted use, you will need to obtain permission directly from the copyright holder.

PART II

The Role of Public Policy in Supporting Sustainable Mobility Transitions

CHAPTER 2

Evaluating Public Policies for Sustainable Mobility: A Review Through Some Interdisciplinary Methodologies and Procedures

Francisco Alonso, Mireia Faus, Sergio A. Useche, José Luis Velarte, and Mónica Alonso

Abstract We recommend acknowledging the importance of evaluation as an undisputable need in developing sustainable mobility policies. To achieve this policy recommendation, we propose to take into account

F. Alonso (✉) · M. Faus · S. A. Useche · J. L. Velarte
INTRAS (University Research Institute on Traffic and Road Safety), University of Valencia, Valencia, Spain
e-mail: francisco.alonso@uv.es

M. Faus
e-mail: mireia.faus@uv.es

S. A. Useche
e-mail: sergio.useche@uv.es

J. L. Velarte
e-mail: jose.l.velarte@uv.es

the following: (1) Evaluations must be comprehensive, multidisciplinary, continuous, summative, rigorous, and economically feasible, led by those responsible for its design and implementation; (2) The incorporation of Key Performance Indicators (KPI) as a structured tool to evaluate the success of policies is recommended; (3) Continuous innovation should be encouraged in policy development and evaluation processes, taking advantage of potential new technological advances to ensure that policies are current, relevant, and effective over time; (4) Emphasize the relevance of involving all stakeholders and incorporating social and community perceptions through different tools and feedback mechanisms; and (5) Conducting a cost–benefit analysis is essential to maximize the effectiveness of budgets that are limited by definition and in reality.

Keywords Benchmarking · Policy assessment · Participatory evaluation

Introduction

Sustainable mobility continues to gain ground and importance in political agendas worldwide, especially in the current context that is characterised by urban growth, sometimes uncontrolled, in which the pollution generated by increased traffic congestion forces the adoption of urgent measures to mitigate the resulting health problems and climate change. These policies seek to transform the way people move around our cities by promoting more efficient and environmentally friendly transport alternatives such as public transport, shared mobility, and soft modes of transport such as cycling and walking. Unfortunately, in many cases, the measures or countermeasures implemented are not subjected to evaluation processes, which implies not knowing the inefficiencies that may occur from an economic, social, and technical point of view. Consequently, adopting rigorous evaluation processes may allow different stakeholders to understand their impact, effectiveness, and efficiency, especially for decision-makers.

M. Alonso
Valencia Provincial Council, Valencia, Spain
e-mail: monicalaura.alonso@dival.es

In addition, the evaluation of sustainable mobility policies is inherently multidisciplinary. To fully understand their impact, it is necessary to address a variety of aspects to be considered, ranging from travel efficiency, including aspects such as the quality of transport itself, infrastructure, its effect on air quality, user perception, equity of access, etc. The disciplines include civil engineering, transportation engineering, logistics, geography, economics, psychology, sociology, data science, and many more.

This chapter will explore how these disciplines converge and complement each other in the policy evaluation process, with the main objective of reviewing and discussing methods of evaluating sustainable mobility measures, with a focus on their applicability and relevance to various contexts. For this purpose, some evaluation methodologies based on scientific literature and European guidelines have been identified, integrating indicators and results through STEM and SSH science procedures to establish a more holistic approach. Therefore, the authors, through consensus and contributing their knowledge of their respective areas of study, have identified four interdisciplinary and potentially valid and applicable evaluation methodologies for most sustainable mobility policies. Subsequently, a review of scientific documents available in relevant and recognised databases such as Web of Science, Scopus, and PubMed has been carried out to synthesize their main findings and recommendations. Inclusion and exclusion criteria were determined to be scientific articles published in Spanish or English.

All the authors of this book chapter, whether from SSH or STEM disciplines, conducted the search and review process independently. Subsequently, the selected articles were pooled and integrated as a result of the SSH-STEM collaboration, thus obtaining a set of methodologies and evidence that integrate their different perspectives on evaluating public policies in this field.

Methods for the Analysis of Public Policies Applicable to Sustainable Mobility Policies

The specialized literature supports the idea that the evaluation of public policies for the development of sustainable mobility is, apart from being helpful, critical to ensure that the strategies, plans, actions, measures, or countermeasures that derive from them are effective and contribute effectively to a more sustainable urban environment. The combination of various methods will provide a comprehensive evaluation of sustainable mobility policies, allowing decision-makers to successively adjust (in a formative rather than summative dynamic) and improve strategies based on the results obtained. In this sense, the following sections present specific factors, and strategies, along with methods to consider when evaluating these policies and systematic actions. The recommendations on evaluation methodologies are potentially applicable at the local, national, or European level, adapting to the characteristics of the policy or countermeasure to be evaluated. The target audience is, therefore, vast, encompassing policymakers and authorities in the mobility sector at different planning levels. In addition, the characteristics of the measure must be taken into account for the selection of the evaluation method, and it is not necessary to apply all of the proposed methods in all cases.

Key Performance Indicators (KPIs)

KPIs are evaluation instruments that provide relevant and specific information on how the established objectives are being achieved. The guidelines for the evaluability assessment of public policies (Institute for the Evaluation of Public Policies, 2020) outline three important points for assessing the results of an intervention: identifying the results, the key mechanisms, and the establishment of indicators. Thus, recent research also points to the effectiveness of KPIs for the evaluation of sustainable mobility policies, providing a framework for measuring and tracking progress towards the United Nations Sustainable Development Goals (SDGs) by allowing the continuous monitoring and evaluation of progress, the identification of areas for improvement and the direction of specific interventions (Quijano et al., 2022). It is recommended that a selection of relevant and representative indicators be made, avoiding a large number of KPIs that reiterate information or do not meet the objectives of the measure.

In this line, and as examples of current applications of KPIs, Hussain et al. (2023) have developed a framework for sustainable mobility and tourism, an indicator-based approach to assess different aspects of the sustainability of smart mobility and tourism projects, concluding how these factors can benefit each other. For their part, Soriano-Gonzalez et al. (2023) examine KPIs related to citizens' mobility and logistics in smart urban areas, identifying environmental data as an area that requires attention in sustainable mobility logistics. Among these KPIs, continuous monitoring and technology monitoring seem to have a major role in effectively evaluating sustainable mobility policies.

The evaluation of sustainable mobility policies through technology addresses two essential perspectives: the use of technological tools to assess the impact of implemented measures and countermeasures and the evaluation of technology as a determining factor in improving sustainable mobility. Hence, technological tools play a crucial role in collecting and analyzing data to assess the impact of specific policies. "Internet of Things" devices and sensors provide real-time information on traffic density, air quality, and the use of different modes of transport, and should be taken into account by policymakers and stakeholders. Among many others, cities such as London, Paris, and Tokyo use camera systems and pattern recognition to monitor traffic flow (Idé et al., 2016). In addition, mobility tracking applications provide data on travel patterns and transportation preferences, facilitating detailed analysis. Using new technologies (mobile systems with real-time cameras and recording with data processing) to collect additional complementary information related to infrastructure characteristics and accidents is widely suggested (Alonso, 2015).

Regarding modelling and simulation tasks, technology makes it possible to predict the future impact of sustainable mobility policies. Using traffic simulation models, it is possible to anticipate the effect of new measures, such as implementing bicycle lanes or pedestrian zones. Predictive analytics algorithms make it possible to forecast changes in mobility and adjust policies proactively (Wei & Mukherjee, 2024). In this way, monitoring is not a static process but allows data to be collected and the results obtained to be evaluated continuously. For example, Ketter et al. (2023) indicate that through connected and autonomous mobility resources, it is possible to obtain preferences of transportation modes in real-time, and to perform analysis and predictive models that allow automated planning and control, being able to anticipate and reduce congestion and favoring a smooth, uninterrupted urban mobility. Moreover, artificial intelligence (AI) is bursting onto the scene. Cirianni et al. (2023) point out that AI can help model alternative mobility system scenarios in real-time (processing big data from heterogeneous sources in a very short time) and identify network and service configurations by comparing phenomena in similar contexts, as well as support the implementation of demand management measures to achieve sustainable objectives.

Evaluating technology as a driver of sustainable mobility focuses on several areas. In the case of electric vehicles, energy efficiency and environmental impact compared to internal combustion vehicles are evaluated (Bi et al., 2016). Carpooling and ridesharing platforms are analysed in terms of their contribution to reducing congestion and emissions by encouraging carpooling. C-ITS (Cooperative Intelligent Transportation Systems) technologies, therefore, play a crucial role. Intelligent traffic management and real-time information systems improve traffic flow and facilitate users' own decision-making. An example of such an evaluation can be seen in the C-Roads Spain Project, whose objective is related to providing a coordinated framework of activity for Spanish stakeholders in the development of effective C-ITS products and services, and which has been developed in a multidisciplinary manner by various entities. Complementarily, the smart use of existing climate intelligence has been claimed to allow informed and more efficient decision-making and increase the resilience of roads, strengthening the adaptation to changing urban needs and ensuring constant progress towards sustainable mobility (Jiménez et al., 2023).

Social Evaluation

It is nowadays a known fact that considering social issues both before and after the implementation of a certain measure or countermeasure, may strengthen sustainable mobility policy outcomes. Among other benefits, it has been shown to facilitate people's commitment and compliance, which can also be increased with other types of actions, such as social and persuasive communication (Faus et al., 2023). For instance, specific measures to deter sexual harassment in public transport, in addition to being evaluated through KPIs, should be complemented with the perspective of users such as women or vulnerable people (Useche et al., 2024). Furthermore, citizen evaluation also allows for identifying the social impact, with a focus on equity, which implies analyzing how these measures and countermeasures affect different groups in society, especially those that may be vulnerable, considering aspects such as accessibility, road safety, or environmental impact, among others (Storme et al., 2021).

Similarly, conducting surveys, public consultations, focus groups, and roundtables (forums) that address aspects such as satisfaction with public transport, perception of infrastructure, social acceptance of the actions developed, or, more generally, the overall evaluation of mobility in the city, offer a valuable perspective on the effectiveness of the measures implemented, as well as positively influencing user behaviour (Whitmarsh et al., 2009). Additionally, these methods are compatible with others by facilitating the community's expression of their opinions, concerns, and suggestions, providing information that is fundamental and differential, and allowing them to simultaneously be involved in the decision-making process (Institute for the Evaluation of Public Policies, 2020). It is highly recommended to use both qualitative and quantitative instruments to have a holistic and complete view of user perceptions.

It is also worth considering that generating communication and publicity campaigns aimed at changing specific attitudes and behaviors to promote sustainable mobility can be highly useful. These communication strategies can improve their effectiveness through knowledge of users' perceptions. Furthermore, it is essential that the campaigns themselves undergo evaluation to measure their effectiveness, as this is an integral part of the broader concept of sustainable mobility. (Faus et al., 2021).

Cost–Benefit Analysis (CBA)

The CBA method evaluates the direct and indirect costs of policy implementation compared to the benefits obtained (Ortuño, 2016). Evaluation through cost–benefit analysis serves the purpose of determining the economic efficiency of sustainable mobility policies. This approach involves quantifying and assessing both the costs and benefits associated with the implementation of specific measures through comparisons between different monetised variables. For example, we could mention the budgetary increase that must be assumed to increase public transport supply versus the reduction of waiting times due to traffic congestion. Another case in point could be the budget allocation needed to install charging stations for electric vehicles versus the health improvement associated with less atmospheric pollution.

Nevertheless, the cost analysis, whose value and effectiveness are rather relative and depend on their rigor and technical quality, should not be carried out only from an institutional or unilateral but participative approach (i.e., from the point of view of how much it will cost an administration to implement a measure) since, in many cases, policies entail restrictions on the movement of citizens, forcing them to modify their mobility habits. As a specific example, the case of Low Emission Zones in urban areas has shown to have a cost for both the government (e.g., installing license plate detection technology) and the residents of the area (e.g., the obligation to purchase less polluting vehicles to access their homes, which can be unaffordable in many cases). Useful previous empirical analyses of this nature are available in Griffin et al. (2020) and Morfeld et al. (2014).

However, despite its widespread use in most European countries during the last decades (Haarich, 2005), recent research indicates that CBA may present deficiencies, especially in complex multi-stakeholder projects or actions as it usually requires that the action is structured as a ranking choice problem with defined decision alternatives, and often this does not match reality, due to existing uncertainties related to the design or impact of the measure (Beukers et al., 2012). For these complex scenarios, Te-Boveldt et al. (2022) recommend replacing or complementing CBAs with social impact assessments, in which multiple stakeholders provide their views on the potential positive or negative impact of the assessed actions. The application of cost-effectiveness analyses is also proposed, a variant of the procedure that is applied when there is a lack of prices to assess the objective to be achieved with the measure (Institute for Evaluation of Public Policies, 2020). In any case, depending on the characteristics of the measure being evaluated, it will be up to the decision-maker to decide whether it is appropriate to use this method or whether it is better to omit it in favor of other forms of evaluation.

Comparison with Best Practices

Today, the baseline of expected quality and effectiveness of public policies is often determined by benchmarking. This concept refers to the task of comparing the evidence and results retrieved from a given context (e.g., a local intervention) with previous or "paradigmatic" experiences with high levels of success, most of them generally collected from pioneering, high-income countries with high maturity in these fields (Garau et al., 2016). It is generally known that this benchmarking process involves analyzing and comparing measures implemented in a particular location with those adopted in other regions that have proven to be effective. However, in some cases it is not known that this requires monitoring local aspects, including obstacles, strengths, opportunities, and potential confounding factors based on user and transport dynamics (e.g., the reasons why people travel, the road user behaviour, environmental friendliness, and safety compliance). Overall, and once these situational factors are considered, the aim is to identify key findings and empirical lessons from previous successful experiences in sustainable mobility to inform and improve local policies (Macmillen & Stead, 2014). Examining success stories provides valuable information on proven effective strategies, allowing for tailoring approaches to the specific needs and conditions of the community.

In addition, benchmarking against best practices focuses not only on outcomes but also on processes and implementing innovative technologies. Knowing how other cities or regions have integrated technological solutions, such as electric vehicles, intelligent transportation systems, or shared mobility initiatives, offers crucial insights. This approach not only helps to optimize the effectiveness of local policies but also contributes to building a global body of knowledge on sustainable mobility, promoting collaboration, and adopting approaches that have proven successful in different contexts. However, and this is a limitation of this coping method, it is necessary to focus on practices that were actually evaluated positively, as unfortunately, too many times, something that is only "practices" are qualified as "good practices", as they lack the required evaluation (Pires et al., 2014). In consequence, it is necessary to establish indicators that endorse effectiveness and allow actions to be categorised as "good practices". In addition, the evaluation can also indicate the "worst practices" and the cause of their low effectiveness, so as not to repeat such measures in other contexts.

Conclusion and Recommendations

Sustainable mobility policies must necessarily include a comprehensive evaluation. Policy developers might benefit from adopting multidisciplinary approaches in the evaluation process. Integrating perspectives from diverse fields leads to a more complete understanding of the impacts of implemented policies. This synergy strengthens the ability to design more effective strategies adapted to the complex interrelationships within a sustainable mobility system.

Key Performance Indicators should be applied for a rigorous evaluation. Incorporating objective, comparable, and verifiable key performance indicators may favour applying, evaluating, and improving sustainable mobility policies. These indicators enable objective performance measurement and facilitate informed and strategic mid- and long-term decision-making.

Innovation and technology facilitating continuous evaluation should be promoted. Fostering an innovative context would enable policies to remain relevant and effective as new developments emerge. This proactive approach, apart from enabling the identification and adoption of emerging solutions gives a dynamic character to sustainable mobility strategies. Moreover, bearing in mind the unique contributions of technological advancements to mobility may further enhance the likelihood of success of these actions.

Considering social (i.e., citizen and collective) perception through valid and mixed research tools may result in feedback, new ideas, and sometimes potential "game-changers". This makes it possible to highlight the importance of considering the experiences and perspectives of users in policy evaluation.

Applying cost–benefit analysis is necessary for an economic evaluation. Any sustainable mobility measure will have a series of costs and benefits that must be integrally and systematically assessed. Moreover, in the balance, not only the direct costs of the measure must be considered, but also the indirect (complementary) costs. All this is particularly important since we must be more than aware that the budgets of public administrations are often limited and demand efficient approaches to maximise their scope. Although the decision to use this method (or others) must take into account its limitations.

References

Alonso, M. L. (2015). *La integración del factor humano en el ámbito técnico de la gestión de las carreteras y la seguridad vial: Un enfoque investigativo* (Doctoral dissertation, Universitat de València, INTRAS).

Alonso, M.L., Parra, L., Jiménez, F., & Crespo, L. (2022). Towards a more resilient Spanish road network. *Routes/Roads, 393*.

Bi, Z., Kan, T., Mi, C. C., Zhang, Y., Zhao, Z., & Keoleian, G. A. (2016). A review of wireless power transfer for electric vehicles: Prospects to enhance sustainable mobility. *Applied Energy, 179*, 413–425. https://doi.org/10.1016/j.apenergy.2016.07.003

Beukers, E., Bertolini, L., & Te Brömmelstroet, M. (2012). Why cost benefit analysis is perceived as a problematic tool for assessment of transport plans: A process perspective. *Transportation Research Part a: Policy and Practice, 46*(1), 68–78. https://doi.org/10.1016/j.tra.2011.09.004

Cirianni, F. M. M., Comi, A., & Quattrone, A. (2023). Mobility control centre and artificial intelligence for sustainable urban districts. *Information, 14*(10), 581. https://doi.org/10.3390/info14100581

Faus, M., Alonso, F., Fernández, C., & Useche, S. A. (2021). Are traffic announcements really effective? A systematic review of evaluations of crash-prevention communication campaigns. *Safety, 7*(4), 66. https://doi.org/10.3390/safety7040066

Faus, M., Fernández, C., Alonso, F., & Useche, S. A. (2023). Different ways… same message? Road safety-targeted communication strategies in Spain over 62 years (1960–2021). *Heliyon, 9*(8). https://doi.org/10.1016/j.heliyon.2023.e18775

Garau, C., Masala, F., & Pinna, F. (2016). Cagliari and smart urban mobility: Analysis and comparison. *Cities, 56*, 35–46. https://doi.org/10.1016/j.cities.2016.02.012

Griffin, S., Walker, S., & Sculpher, M. (2020). Distributional cost effectiveness analysis of West Yorkshire low emission zone policies. *Health Economics, 29*(5), 567–579. https://doi.org/10.1002/hec.4003

Haarich, S. N. (2005). Diferentes sistemas de evaluación de las políticas públicas en Europa: España, Alemania y los países del Este. *Revista Española De Control Externo, 7*(20), 64–88.

Hussain, S., Ahonen, V., Karasu, T., & Leviäkangas, P. (2023). Sustainability of smart rural mobility and tourism: A key performance indicators-based approach. *Technology in Society, 74*, 102287. https://doi.org/10.1016/j.techsoc.2023.102287

Idé, T., Katsuki, T., Morimura, T., & Morris, R. (2016). City-wide traffic flow estimation from a limited number of low-quality cameras. *IEEE Transactions on Intelligent Transportation Systems, 18*(4), 950–959.

Institute for the Evaluation of Public Policies. (2020). *Guidelines for the evaluability assessment of public policies*. Spanish Government, Spain.

Jiménez, F., Crespo, L., Gil, A., Parra, L., Alonso, M. L., & Collazos, F. (2023). *Risk thresholds related to climate change in road infrastructure in Spain*. Prague 2023 XXVIIth World Road Congress. https://www.road.or.jp/english/img/piarc/Climatechangeandresilience.pdf

Ketter, W., Schroer, K., & Valogianni, K. (2023). Information systems research for smart sustainable mobility: A framework and call for action. *Information Systems Research, 34*(3), 1045–1065. https://doi.org/10.1287/isre.2022.1167

Macmillen, J., & Stead, D. (2014). Learning heuristic or political rhetoric? Sustainable mobility and the functions of 'best practice.' *Transport Policy, 35*, 79–87. https://doi.org/10.1016/j.tranpol.2014.05.017

Morfeld, P., Groneberg, D. A., & Spallek, M. F. (2014). Effectiveness of low emission zones: Large scale analysis of changes in environmental NO_2, NO and NOx concentrations in 17 German cities. *PLoS ONE, 9*(8), e102999. https://doi.org/10.1371/journal.pone.0102999

Ortuño, A. (2016). Diagnóstico y propuestas para una adecuada planificación de infraestructuras en España. *Revista De Obras Públicas, 3575*, 71–78.

Pires, S. M., Fidélis, T., & Ramos, T. B. (2014). Measuring and comparing local sustainable development through common indicators: Constraints and achievements in practice. *Cities, 39*, 1–9. https://doi.org/10.1016/j.cities.2014.02.003

Quijano, A., Hernández, J. L., Nouaille, P., Virtanen, M., Sánchez-Sarachu, B., Pardo-Bosch, F., & Knieilng, J. (2022). Towards sustainable and smart cities: Replicable and KPI-driven evaluation framework. *Buildings, 12*(2), 233. https://doi.org/10.3390/buildings12020233

Soriano-Gonzalez, R., Perez-Bernabeu, E., Ahsini, Y., Carracedo, P., Camacho, A., & Juan, A. A. (2023). Analyzing key performance indicators for mobility logistics in smart and sustainable cities: A case study centered on Barcelona. *Logistics, 7*(4), 75. https://doi.org/10.3390/logistics7040075

Storme, T., Casier, C., Azadi, H., & Witlox, F. (2021). Impact assessments of new mobility services: A critical review. *Sustainability, 13*(6), 3074. https://doi.org/10.3390/su13063074

Te-Boveldt, G., Keseru, I., & Macharis, C. (2022). When monetarisation and ranking are not appropriate. A novel stakeholder-based appraisal method. *Transportation Research Part A: Policy and Practice, 156*, 192–205. https://doi.org/10.1016/j.tra.2021.12.004

Useche, S. A., Colomer, N., Alonso, F., & Faus, M. (2024). Invasion of privacy or structural violence? Harassment against women in public transport environments: A systematic review. *PLoS ONE, 19*(2), e0296830. https://doi.org/10.1371/journal.pone.0296830

Wei, Z., & Mukherjee, S. (2024). Analyzing and forecasting service demands using human mobility data: A two-stage predictive framework with decomposition and multivariate analysis. *Expert Systems with Applications, 238*, 121698. https://doi.org/10.1016/j.eswa.2023.121698

Whitmarsh, L., Swartling, Å. G., & Jäger, J. (2009). Participation of experts and non-experts in a sustainability assessment of mobility. *Environmental Policy and Governance, 19*(4), 232–250. https://doi.org/10.1002/eet.513

Open Access This chapter is licensed under the terms of the Creative Commons Attribution 4.0 International License (http://creativecommons.org/licenses/by/4.0/), which permits use, sharing, adaptation, distribution and reproduction in any medium or format, as long as you give appropriate credit to the original author(s) and the source, provide a link to the Creative Commons license and indicate if changes were made.

The images or other third party material in this chapter are included in the chapter's Creative Commons license, unless indicated otherwise in a credit line to the material. If material is not included in the chapter's Creative Commons license and your intended use is not permitted by statutory regulation or exceeds the permitted use, you will need to obtain permission directly from the copyright holder.

CHAPTER 3

Enhancing Policy Coordination of Schooling and Transport for Net-Zero

Joshua Lait, *Femke Nijsse*, *Stewart Barr*, *David Hall*, *Catherine Butler*, *Antonio Olmedo*, *and Cormac Lynch*

Abstract We recommend enhancing policy coordination of schooling and transport for achieving Net-Zero. To achieve this policy recommendation, we propose to take into account the following: (1) Education policy can adversely impact sustainable transport transitions; (2) Interactions between schooling and transport do not feature in policy and strategy; (3) Existing datasets fail to capture schooling as a component of the transport system; (4) Coordinating data creation and collation

J. Lait (✉) · F. Nijsse · S. Barr · D. Hall · C. Butler · A. Olmedo · C. Lynch
University of Exeter, Exeter, UK
e-mail: j.lait@exeter.ac.uk

F. Nijsse
e-mail: f.j.m.m.nijsse@exeter.ac.uk

S. Barr
e-mail: s.w.barr@exeter.ac.uk

D. Hall
e-mail: d.j.hall@exeter.ac.uk

© The Author(s) 2024
I. Keseru et al. (eds.), *Strengthening European Mobility Policy*,
https://doi.org/10.1007/978-3-031-67936-0_3

can enhance understandings of different sectors' impacts on transport systems; and (5) Scaling-up existing areas of agreement over priorities and cooperation between actors can help to support sustainable transport transitions.

Keywords Policy conflict · Education policy · Cross-sectoral impacts

Introduction

Significant energy demand reduction is necessary over the short and medium terms to meet net-zero commitments and address climate change (Anderson et al., 2014). The promotion of incremental reductions through technological innovations (such as increased efficiency) or market policies (such as price mechanisms) is unlikely to be sufficient in delivering the deeper changes needed to support sustainable transitions (Pearse & Böhm, 2015). In this chapter, we recommend enhancing the cross-sectoral coordination of national policy objectives, governance structures, and datasets as a radical intervention for reducing energy demand in the Europe Union, which is a core dimension of *REPowerEU* (2022) and the *Energy Efficiency* directive (2023).

We define "policy coordination" as an environmental governance approach that enhances cross-sectoral collaboration and cooperation between actors during the policymaking process to mitigate the perverse impacts of siloed decision-making (Jordan & Lenschow, 2010). Coordination at a system scale is crucial because policies that are not intended to impact energy demand can pose important, but often unrecognised impacts (Royston et al., 2018). Here, we focus on the need for enhancing

---

C. Butler
e-mail: c.butler@exeter.ac.uk

A. Olmedo
e-mail: a.olmedo@exeter.ac.uk

C. Lynch
e-mail: cl852@exeter.ac.uk

coordination in EU policymaking through the example of the interactions between schooling and travel. Education policies that allow for school marketisation and parental choice of an institution within a market of schools are significant to transport governance because they permit children to attend the school that is not the closest.

Unlike higher education, primary and secondary education levels in Europe remain under the complete control of each individual member state. Therefore, the extent of school marketisation and choice policies has varied within and between different EU nations (Gunter et al., 2016; Kresjler & Moos, 2023). This process remains best described as quasi-marketisation (le Grand & Bartlett, 1993) whereby various dimensions of schooling are opened up to intense marketisation while the core financing of schools remains subject to public spending. Within the varying deployment of (quasi)market education policies in Europe, school choice is perhaps the most widely enacted principle. Despite varying levels of regulation in each European country (Dronkers & Avram, 2014), families increasingly perceive that they have a right to choose "the" individual school that they prefer for their children and, as a result, can operate strategically within the different national regulatory settings to achieve their goal (see Dupriez & Maroy, 2003; Olmedo, 2008).

The inclusion of (quasi)market and choice policies as part of a suite of school reforms has nevertheless been widespread, perhaps, nowhere more so than in England (Hall, 2023). Goals that relate to parental choice have been key to the process by which children are allocated to schools in this context. The liberalisation of schooling and deregulation of transport provision occurred during the nation's membership of the EU, suggesting that the challenges posed by this eventuality could also occur in other European nations. As an extreme case of education system liberalisation, the transport implications of the case study are of considerable interest to national and regional policymakers in European nations that are pursuing these policies.

To understand these cross-sectoral intersections, we developed an interdisciplinary methodology that was informed by insights from the environmental social sciences, sociology of education policy, and complexity science. We adopted the same approach, using different social science and systems modelling methods, to identify, describe and understand how "school travel" in the South West England features in transport datasets and documentary evidence. We aimed to identify how existing datasets do or do not capture the impacts of schooling specifically, highlighting the different types of potential interactions between schooling and other components of local or regional transport systems. We also used social science insights to examine how school travel does or does not feature in different levels of education, transport, and sustainability policy. We explored important interactions and gaps through semi-structured interviews with stakeholders in local government, school leadership, an NGO, a transport company, and parents.

The mixed methodological approach included a) scoping analysis of transport datasets (61 total), b) scoping analysis of policies and documents (45 total), and c) semi-structured stakeholder interviews (15 total) (ZENODO). The research was conducted between November 2023 and March 2024.

Results

Our key findings are summarised below:

- Policy conflicts can undermine sustainable transitions.
- Cross-sectoral impacts are not well recognised in policy and strategy.
- It is difficult to identify and understand the cross-sectoral impacts of schooling within current systems of data creation.
- Liberalisation of schooling and local transport has resulted in a proliferation of actors that are in need of coordinating.

We found that conflicts between sectoral strategies can significantly undermine sustainable transitions, revealing a need for greater cross-sectoral coordination. Firstly, we discuss key examples from our research that highlight how such conflicts are playing out in the context of schooling and transport. An example of such a challenge includes the interactions between changing school funding mechanisms, pupil recruitment practices, and transport patterns. Prior to the 1980s, national funds were distributed in a locally devolved system first to local authorities and then from local authorities to individual schools with considerable discretion at the local level in relation to the distribution of these funds between schools. Following the local financial management of schools initiative in the 1980s, schools began to receive their funding directly based upon a nationally agreed model of formula-based funding linked primarily to levels of student recruitment at each individual institution. This acted to incentivise student recruitment by individual schools given the financial rewards associated with maintaining or expanding the numbers of students on roll. This process of "incentivisation" to recruit became ever more urgent as schools which experienced significant decreases in student recruitment found themselves variously closed, taken over, or subject to mergers. We found that parental flight, and in extreme cases, school closure, can result in longer and more costly commutes to attend schools with better reputations, which conflicted with goals promoting sustainable, affordable, and active travel.

Secondly, the analysis also showed that market reforms of education in England actively encourage parents to choose schools based on reputation, which can result in pupils travelling further distances by additional bus service or car, and at greater expense, to local authorities, schools, and parents. The creation of performance league tables, along with reports and rating systems have all served to offer reputational information upon which parents are encouraged to make "choices" as to which school their child(ren) should attend. One important implication of this for travel is that parents widely view themselves as active choosers in deciding upon the school that their child(ren) can attend and can decide against the nearest or local school if it is considered to have a poor reputation. This capacity to choose non-local schools intersects with decreasing coverage of public transport routes in the South West, contributing to car reliance at the primary education level where parents typically escort children on their journey. School commutes over longer distances can inhibit active travel as a mode choice at all school levels, undermining health and sustainability goals.

We found that this capacity to choose non-local schools places additional pressures on local authority transport officers that oversee the financing, planning, and operation of travel networks to meet these additional travel needs, which change annually with pupil intake and are often greatest for rural communities. Such incoherence is particularly evident in the cases of pupils with special educational needs and disabilities (SEND) or those attending selective schools. SEND education is provided in a smaller range of separate sites to local mainstream schools. The choice of SEND school, which is made in collaboration between parent and local authority, can require pupils with SEND to commute even greater distances than those that attend mainstream schools. This can produce substantial transport costs for a local authority. Likewise, selective schools often rely on bussing to ensure that pupils attend from a wide range of communities sometimes far beyond a school's locality. Schools and parents pay the costs of bussing. Ultimately, these outcomes conflict with national and regional transport strategies that promote more sustainable and affordable forms of daily travel.

Unfortunately, the cross-sectoral impacts of education liberalisation are not well recognised in policies and documents at all scales. The fragmented nature of policymaking means that most policies and documents sampled only included brief references to school travel. This includes the aims to promote school street closures during drop off and pick up times or increasing EV charging capacity on schools in national transport strategy and a duty for local authorities to promote sustainable school commuting in national education guidance. The sampled regional admissions policy and local school attendance documents contained brief mentions of transport. Certain types of school-related travel were distinctly absent in all documents, such as travel relating to SEND learning, extracurricular activities, and community use of school premises.

Importantly, the inclusion of practical "school travel" interventions decreased at the local authority and even further at the school scale. This suggests that there is a coherence challenge emerging between higher, regional, and local governance levels. For example, regional transport and air quality plans state aims to link schools as well as other important locations using cycling/walking networks, without considering how travel to non-local schools can make active travel impossible or how schools should be engaged in this process. No schools sampled published specific or up-to-date transport policies. However, we found that school policies, relating to pupil recruitment, admissions, and operational hours can play a role in affecting the patterns and timings of school commutes, posing implications for levels of congestion and related air pollution near schools. Yet, these impacts typically do not feature within policy assemblages relating to areas like admissions, attendance, and exclusions. Overall, these cases highlight the importance of considering *how* to enhance our understanding of the unintended consequences of different objectives and goals to more effectively coordinate governance processes across sectors.

Similarly, the fragmented nature of the available data means that only a partial understanding of school-related transport data can be obtained via existing datasets. Available data sources are not designed specifically with schooling in mind. This means that existing datasets of local and regional transport fail to capture schooling as a component of the system.

For example, available data reveals little about travel behaviour, which is necessary to understand how different policies (at the local authority or school level) influence travel choices. Consequently, we can draw no conclusions from the datasets reviewed about the travel mode split and how this is reflected spatially and temporally in local authority areas, or indeed in terms of how different approaches adopted by schools may influence travel practices. In other words, there is a distinct absence of available behavioural data that helps us to understand how different policies (at the local authority or school level) influence travel choices.

The resolution of datasets reviewed was limited to broad levels (e.g., local authority) and so did not have the granularity required to understand school travel at appropriate scales (e.g., Middle/Lower Layer Super Output Areas). A relevant example is how school demand and capacity are projected to assess transport provision needs. Both the Department of Education and Department for Transport datasets we reviewed did not have the level of granularity required to understand transport demand from schools at appropriate scales, such as Middle/Lower Layer Super Output Areas (as opposed to the whole local authority area). Therefore, it is difficult to make sense of the potential effects of liberalisation on local/regional transport patterns in the data sampled.

There are often gaps in existing data that could be addressed through coordination. For example, the Bus Open Data Service provides school bus and bus stop information only for major bus companies. Distinguishing between the percentage of pupils using dedicated school buses or using public bus services (which might include interchanges) is also difficult with existing data. There is also a lack of available data on congestion around schools. Further, whilst data on key demographic factors was available to understand access to school transport at a national scale (including ethnicity, income decile, and residence (urban/rural)), this data was not available at the local authority level. This raises questions about how easily national and regional policymakers can understand school transport patterns and inequalities or how to tackle these through coordination at regional and local levels.

The liberalisation of education and transport has led to the emergence of a range of different actors with distinct roles that are in need of coordinating. The diversification of school type in England, which has strongly complemented the choice process, has resulted in a proliferation of school types according to gender mix, religion, academic selectivity, or another of the many methods of school differentiation in this context (Courtney,

2015). For example, an "academy" is a new type of publicly funded school that is privately-run through a Trust rather than a local authority, who was traditionally responsible for overseeing school and transport governance. This partitioning can contribute to new coordination challenges, such as the priority conflicts between a school's pupil recruitment ambitions and a local authority's desire to reduce transport costs for busing pupils to schools that are not the nearest. We also found that transport deregulation has compounded this problem by increasing the number of national, regional, and local private transport companies providing public and/or dedicated school bus services that local authorities orchestrate at great expense in travel networks, which change annually in line with school allocations. Overall, this emphasises the need for a greater bundling of responsibilities at the regional and local policy levels to govern these disparate networks of privately-run stakeholders more effectively.

Lastly, stakeholders with distinct and bounded roles often do not recognise how national education policies and local school policies can affect local/regional transport patterns. For example, school leaders perceived transport as primarily the responsibility of parents and pupils, over which they have little to no control. Few of the schools sampled published a specific transport policy. While stakeholders working at the regional scale in a local authority and an NGO discussed a range of small-scale travel interventions, such as promoting behaviour change, there was generally less recognition of the impacts of broader education reforms on local/regional transport patterns. Notably, stakeholders reported how teams responsible for school allocations at the local authority have typically worked separately from those responsible for planning and provisioning the resultant transport routes. This demonstrates a need to enhance the coordination of actors and policies across different non-transport decision-making levels to address such perverse outcomes.

Conversely, local authority stakeholders reported that escalating transport costs have led to transport officers participating in allocation meetings to advise on the potential costs of particular school admission decisions. This is an example of informal mutually beneficial coordination, whereby particular shared values or understandings (such as reducing costs, promoting active lifestyles, ensuring safety, and promoting environmental sustainability) can encourage stakeholders with different roles and expectations to address a policy problem through collaboration. This emerging response signals the potential role policy coordination of actors could play in mitigating the challenges posed by the cross-sectoral policy conflicts identified in this research.

Discussion

There is a need to coordinate different objectives and goals in order to better understand interactions and conflicts. Specifically, we have drawn attention to the transport challenges posed by the complex overlaying of liberalisation policies in the education sector in England. This has involved attempts to "improve" schools and the school system more generally by unleashing market forces, choice, and competition within a financial model underpinned by continued state funding. Goals that relate to parental choice have been key to the process by which children are allocated to schools. We found that the capacity to choose the school a pupil attends in a (quasi)market is highly significant to travel governance because it permits pupils to attend the school that is not the closest, resulting in challenges for promoting active travel and reducing car dependency for school commuting. These inadvertent outcomes of liberalisation conflict with transport policy goals evident in England and in the EU's *Sustainable and Smart Mobility Strategy*: to promote healthy and sustainable local/regional travel. In addition, the analysis revealed that attendance and exclusions policies can affect travel timings and needs, posing inadvertent implications for levels of congestion-related pollution near schools. Similarly, this outcome can conflict with the EU's *Zero Pollution* policy targets (2021) for reducing healthy inequalities by improving air quality.

We argue that policy coordination could be engendered, then, in part through attention to the data required to build an understanding of how policy goals and their effects intersect. To this end, we suggest that local and regional authorities in Europe must begin to combine disparate data sets, using situated knowledge of data creation in each context, to begin to build a picture of the effects of how policy goals intersect (and conflict). This could involve local transport authorities compiling datasets produced by a range of organisations that are either available as open source or obtainable on request. Orchestrating data across levels can help to develop a deeper understanding of the transport implications of multilevel policy interactions. We found that necessary data are often fragmented and not publicly available. Yet, stakeholders report that these data are often held by schools that are evaluating interventions to change behaviour, and we believe are also held by private transport providers and phone companies who do not share this data freely. Ultimately, we believe that national and regional governments in the EU need to coordinate data sources more effectively to deepen understanding of cross-sectoral interactions and to identify interventions that align with the EU policy goals identified in this chapter.

Finally, the liberalisation of education and transport provision at the local/regional scales has led to a proliferation of actors that are in need of coordination to support sustainable transitions. Thus, regional and local governing bodies are needed to orchestrate actors engaged in different areas of public policy (such as regional and urban planning, transport authorities, school governance, and parental organisations) but also new/renewed actors (such as parent choosers/consumers, private transport providers, and housing developers). The authors recognise that there are potential limitations to coordination (such as the additional costs and administrative burden of orchestrating actors) and that distinct and bounded roles are often necessary from an operational perspective to support targeted decision-making. As a result, there are undoubtedly limits of the extent to which it is possible or indeed desirable to coordinate multiple actors in different sectors. Nevertheless, we argue that scaling-up existing areas of priority agreement and informal collaboration between stakeholders, implicated in incoherence challenges linked to liberalisation, through formalising relationships and bundling governance powers in local/regional government bodies represents a crucial pathway for enhancing policy coordination.

Conclusion and Recommendations

In this chapter, we show how education reforms promoting marketisation and choice that are unfurling across the EU can influence local/regional transport systems. We argue that parental choice is significant for transport governance because it permits pupils to attend the school that is not the closest. This demonstrates a need to coordinate policy and strategy beyond the transport sector to support sustainable transport transitions.

We undertook scoping analyses of cross-sectoral interactions using an interdisciplinary research approach, which was informed by insights and methods from the social sciences and complexity science. The analysis of documents and interviews identified a range of interactions between schooling and transport. This included pupils travelling at greater distances and at greater expense to schools with better reputations, which is particularly significant for those attending SEND or selective schools. We also found that school policies relating to admissions, operational hours, attendance, and exclusions can influence local/regional transport patterns. Importantly, such interactions are not well recognised in existing policy and strategy.

Existing transport datasets fail to capture the possible effects of schooling on the transport system. We found that existing datasets do not account for dedicated school bussing, congestion near schools, and demographic factors necessary to understanding access to school transport at the local authority level. The fragmented nature of available data means that we can only develop partial understanding of schooling as a component of the transport system.

We recommend that policy coordination could be engendered through attention to the data required to build understanding of how policy goals and their effects on transport systems intersect. These additional requirements include combining disparate datasets, addressing key gaps (such as congestion near schools), and improving levels of granularity to understand transport demand from schools. We suggest that national departments and regional governments of EU nations need to orchestrate disparate data sources to enhance understandings and identify interventions that align with transport and other policy goals.

We recommend two avenues for enhancing coordination in governance. First, national and regional policymakers need to identify and formalise areas of existing alignment between objectives and goals in different sectors (such as school access and active, affordable, and sustainable travel) through the creation of cross-sectoral strategies. Second, regional and local authorities need to formalise and scale-up existing areas of informal collaboration (such as factoring transport costs into school allocation decisions) through the bundling of powers and stakeholders' voices in strategic arrangements, which bring together the governance of allocations, school siting, transport funding, housing, and infrastructure development, for each of its local areas.

REFERENCES

Anderson, K., Le Quéré, C., & Mclachlan, C. (2014). Radical emission reductions: The role of demand reductions in accelerating full decarbonization. *Carbon Management*, 5(4), 321–323. https://doi.org/10.1080/17583004.2014.1055080

Courtney, S. J. (2015). Mapping school types in England. *Oxford Review of Education*, 41(6), 799–818. https://doi.org/10.1080/03054985.2015.1121141

Dronkers, J., & Avram, S. (2014). What can international comparisons teach us about school choice and non-governmental schools in Europe? *Comparative Education*, 51(1), 118–132. https://doi.org/10.1080/03050068.2014.935583

Dupriez, V., & Maroy, C. (2003). Regulation in school systems: A theoretical analysis of the structural framework of the school system in French-speaking Belgium. *Journal of Education Policy*, 18(4), 375–392. https://doi.org/10.1080/0268093032000106839

Gunter, H., Grimaldi, E., Hall, D., & Serpieri, R. (Eds.). (2016). *New Public Management and the Reform of Education: European Lessons for Policy and Practice*. Routledge.

Hall, D. (2023). England: Neo-liberalism, regulation and populism in the educational reform laboratory. In J. Kresjler & L. Moos (Eds.), *School policy reforms in Europe*. Springer.

Kresjler, J., & Moos, L. (2023). *School policy reforms in Europe*. Springer.

Le Grand, J., & Bartlett, W. (1993). *Quasi-markets and social policy*. Macmillan.

Olmedo, A. (2008). Middle-class families and school choice: Freedom versus equity in the context of a "local education market." *European Educational Research Journal*, 7(2), 176–194. https://doi.org/10.2304/eerj.2008.7.2.176

Pearse, R., & Böhm, S. (2015). Ten reasons why carbon markets will not bring about radical emissions reduction. *Carbon Management*, 5(4), 325–337. https://doi.org/10.1080/17583004.2014.990679

Royston, S., Selby, J., & Shove, E. (2018). Invisible energy policies: A new agenda for energy demand reduction. *Energy Policy*, 123, 127–135. https://doi.org/10.1016/j.enpol.2018.08.052

Open Access This chapter is licensed under the terms of the Creative Commons Attribution 4.0 International License (http://creativecommons.org/licenses/by/4.0/), which permits use, sharing, adaptation, distribution and reproduction in any medium or format, as long as you give appropriate credit to the original author(s) and the source, provide a link to the Creative Commons license and indicate if changes were made.

The images or other third party material in this chapter are included in the chapter's Creative Commons license, unless indicated otherwise in a credit line to the material. If material is not included in the chapter's Creative Commons license and your intended use is not permitted by statutory regulation or exceeds the permitted use, you will need to obtain permission directly from the copyright holder.

CHAPTER 4

Improving Social Justice, Environmental Integrity, and Geopolitical Resilience in EU Electric Mobility Transition

*Aleksandra Lis-Plesińska, Nathalie Ortar,
Rafał Szymanowski, Marek W. Jaskólski,
Aleksandra Parteka, and Christine Buisson*

Abstract We recommend improving social justice, environmental integrity, and geopolitical resilience in electric mobility transition. To achieve this policy recommendation, we propose the following: (1)

A. Lis-Plesińska (✉)
Department of Anthropology and Ethnology, Adam Mickiewicz University, Poznań, Poland
e-mail: alis@amu.edu.pl

N. Ortar
ENTPE-University, Vaulx-en-Velin, France
e-mail: nathalie.ortar@entpe.fr

R. Szymanowski
Faculty of Political Science and Journalism, Adam Mickiewicz University, Poznań, Poland
e-mail: rafal.szymanowski@amu.edu.pl

Increase societal acceptance and justice of climate policies by engaging local stakeholders; (2) Prioritize sustainable mobility practices over replacement of internal combustion engine vehicle (ICEV) with battery electric vehicle (BEV); (3) Base resilience of global value and supply chains on a diversified network of suppliers and a balanced structure of domestic and foreign content of economic value; (4) Evaluate geopolitical risks and environmental impacts of value and supply chains in non-European regions; (5) Create a geopolitical risk body to scrutinise geopolitical threats to the electric mobility supply chain; and (6) Increase the share of EU-based manufacturing in electric mobility related sectors.

Keywords Social acceptance · Stakeholder engagement · Geopolitical risks

Introduction

The relevance of the electrification of individual vehicles as a solution to reaching ambitious CO_2 reduction objectives needs to be problematized. This chapter proposes to rethink the EU electric mobility transition frequently conceptualized as a shift from ICEV to BEV without other solutions being considered, the importance of spatial considerations being overlooked, and the effects on local and international territories being disregarded.

M. W. Jaskólski
AMU Research Centre for Energy and Environmental Sciences, Adam Mickiewicz University, Poznań, Poland
e-mail: marek.jaskolski@amu.edu.pl

A. Parteka
Faculty of Management and Economics, Gdansk University of Technology, Gdańsk, Poland
e-mail: aparteka@zie.pg.edu.pl

C. Buisson
Gustave Eiffel University, Lyon, France
e-mail: christine.buisson@univ-eiffel.fr

In this chapter, an interdisciplinary team of researchers proposes recommendations with regard to strengthening societal, economic, environmental, and geopolitical resilience of European climate policies in automotive industry and transport systems. Based on a number of research projects,[1] these threats have been identified as resulting in the lack of **coherence** of EU policies, potential **ruptures**, and overlooked **multiplicity** of factors and contexts that might give way to inappropriate development of electric mobility.

Questions regarding EU's policy **coherence** for shaping electric mobility transition emerge as it becomes clear that the CEE Member States are often lagging behind the Northern and Western European frontrunners, as for example with regard to BEVs charging infrastructure (gridX, 2023). Scholars also critically reflect on the weakening of the environmental integrity of electric mobility policies by pointing out highly exploitative practices of extracting climate change commodities in other regions, for example, lithium (Dorn et al., 2022).

In the case of BEV, a growing number of studies (Flipo et al., 2023; Ortar & Ryghaug, 2019) reveal **ruptures** between urban and rural areas as the first ones to adopt e-mobility solutions faster. Within the multilevel structure of EU governance and its policy processes based on common targets, there is always a potential threat of overlooking the **multiplicity** of ways in which European economies, or regions, can decarbonize.

The projects that each author collaborated on were interdisciplinary, although they did not necessarily cross the boundaries between STEM and SSH. In this chapter, we joined forces to arrive at a better understanding of the causal role social innovation plays in shaping, accelerating, or decelerating change trajectories.

The overarching theoretical framework that we adopt for bringing together STEM and SSH perspectives is socio-technical transitions. Irrespective of the exact model used to explain how socio-technical transitions develop, the mixed socio-technical nature of these processes is theorized, and the need for interdisciplinary methodologies underlined. Transitions are thus determined both by materialities of technologies as well as politics, cultures, habits, practices, and the ways in which gains and losses are calculated.

Guided by these insights we: ethically problematize the engineering methodologies for measuring climate impacts of BEVs and ICEVs in life cycle analysis, geopolitically problematize the economic measures of the robustness of Global Value Chains, and socio-culturally problematize pan-European targets by pointing to diverse local outcomes and idiosyncrasies.

To substantiate our approach further, in this chapter, we problematize the European mobility culture of private car ownership and contribute with STEM knowledge to an argument for radically different patterns of mobility and for an ethical reflection on environmental regulatory ruptures between the upstream and downstream of the BEV supply chains. Consequently, we discuss supply chain resilience measures developed in economics at the background of the changing geopolitical landscape.

Applying STEM and SSH expertise together, we highlight the potential threat of overlooking negative environmental, societal, geopolitical, and economic impacts of electric mobility development understood mostly as a replacement of ICEVs with BEVs. STEM knowledge develops, questions, and improves methodologies for measuring and calculating environmental impacts or life cycles of particular technologies. With SSH knowledge one is able to interpret the social, geographic, and political consequences of these methodological choices at different scales: for EU, national politics and locally. Therefore, we see our main contribution in presenting and critically discussing different methodologies and approaches used for generating data and evidence for policymaking, and not discussing the data as such.

The Landscape of EU Electric Mobility Policies

The EU managed to champion itself as a global climate action leader in the early 2000s, implementing the European Emission Trading Scheme and many stringent environmental regulations (Bradford, 2020). On April 17, 2019, the European Commission proposed legislative amendments to Regulation (EU) 2019/631 of the European Parliament and the Council to elevate emission standards for passenger and light commercial vehicles to be more ambitious by 2030, alongside establishing a target for zero emission vehicles by 2035. These proposals align with the objectives of the European Climate Law, which seeks to achieve EU climate neutrality by 2050 and reduce emissions by 55% by 2030, relative to 1990 levels. The EU's recent European Green Deal involves seven actions within the "Sustainable and Smart Mobility" section aiming to support sustainable and smart mobility, deployment of public recharging and refuelling points, and boosting the production and supply of alternative fuels.

Since the introduction of the EU's European Green Deal in December 2020, a gradual shift from a central focus on climate issues to a wider program of economic reform has occurred, as a result of a triple shock of the Covid-19 pandemic, energy and supply chain crisis, and Russia's war against Ukraine. The policy responses to these crises have reinforced some of the goals of the EU's European Green Deal, e.g. energy transition and circular economy, but relaxed others, including rules about pesticide use and agricultural emissions. President of the EC Ursula Von der Leyen was open about this shift in her 2023 State of the Union Speech, admitting that "We shifted the climate agenda to being an *economic* one". The question of how to reconcile its many conflicting aims and decouple economic growth from resource use thus becomes vital.

The project of electromobility transition as one of the pillars of Fit for 55 package includes an extended set of emission reduction methods for road transport:

- Establishing an emissions trading system (ETS2) for the road transport sector operational in 2027.
- Revising the Energy Taxation Directive to remove any tax advantages associated with the utilisation of fossil fuels.
- Enhancing efforts within the Alternative Fuels Infrastructure Regulation to advance its development and boost the deployment of zero- and low-emission vehicles.

EVIDENCE BASE: A CRITICAL DISCUSSION OF THE EXISTING METHODOLOGIES AND APPROACHES

The Ethical Questions of ICEVs and BEVs

Environmental impact assessment methods allow to compare various individual vehicle options. First developed in the 1990s, life cycle analysis (LCA) methods are typically "based on a systematic examination of the environmental impacts of products/activities, with the aim of revealing the environmental dimension of sustainability" (Goedkoop et al., 2008). Many studies have presented comparative analyses of ICEV and BEV during their entire life cycle, from building to recycling, with an analysis of the usage impacts. Depending on the study, the identified effects varied, but all comprise an analysis of the CO_2 impacts. All studies (see for example Tagliaferri et al., 2016) have revealed that the order of magnitude of the CO_2 impact of BEV is not different from that of internal combustion engine vehicles. Indeed, the reported difference is a factor of 2 to 4 in terms of CO_2 emissions. Additionally, the electricity production mix has a strong impact on the global CO_2 emission during vehicle usage. We need to achieve reductions of CO_2 emissions of individual cars by a factor larger than 4, if we want to reach our reduction objectives. Therefore, we need not only to study the path towards replacing all the ICEV with BEV but also to change the mobility patterns.

Looking at the emissions performance of BEV beyond the use cycle, sheds new light on their environmental impacts. ICEV pollute European cities mainly due to PM and NOx emissions when driving. In contrast, BEV primarily pollute during the extraction and refining of the minerals needed to make the batteries, mainly in developing countries e.g. Congo DRC, Zimbabwe, or Argentina (Lèbre et al., 2020). A recent study shows that greenhouse gas emissions generated during the production of BEVs, especially related to the powertrain, are higher than for an ICEV. For a BEV with an 82 kW/h battery, this is 10.12 t CO_2 eq. For an ICEV with a petrol engine, it is only 1.21 t CO_2 eq. At 83% (8.37 tonnes of CO_2 eq.), battery production is the main cause of the high CO_2 emissions during production (VDI, 2023).

Replacing ICEVs with BEVs on a one-to-one basis may not be feasible with the capabilities of current technology and available resources, the replacing would help to reduce the tailpipe emissions, but it does not address issues such as congestion and inefficient land use (Henderson, 2020). Therefore, rethinking our approach to mobility should include not only changing propulsion systems, but also reconsidering the way we move around cities. Investing in and promoting alternatives to individual car use, such as public transport, micromobility solutions, and mobility-as-a-service (MAAS) platforms, can more significantly decarbonise as well as alleviate congestion, reduce the need for vehicle ownership and increase accessibility (Alyavina et al., 2020). The ethical implications of pollution outsourcing also call for rethinking our approach to electric mobility as a tool to reduce GHG emissions (Dorn et al., 2022; Sovacool et al., 2020).

Making Supply Chains More Resilient to Economic and Geopolitical Impacts

The dependence of the European Union on imports of key raw materials such as rare earths, lithium, and cobalt is becoming a major challenge (Bolitho, 2023; European Commission, 2020, 2023). In March 2024, the European Council adopted the European Critical Raw Materials Act, as demand for rare earth elements is expected to grow exponentially in the coming years.

The act is expected to:

- increase and diversify the supply of critical raw materials in the EU
- strengthen closed-loop cycles, including recycling
- support research and innovation on resource efficiency and substitutes.

The electric mobility sector is characterised by strong cross-country dependencies within global value chains/global supply chains (GVCs/GSCc). Indicators of GVC linkages based on input-output data (Antràs & Chor, 2022) help to trace where the economic value is created, assess the reliance on foreign inputs and quantify the magnitude of insecurity faced by producers, consumers, and policymakers (Baldwin & Freeman, 2022). The resilience of specific sectors, such as the electromobility sector, result from the insufficient supply chain diversification, tight structure of logistic network (bottlenecks), concentrated sourcing of raw materials, components, and inputs, as well as risks of labour supply shortages. The quantification of the adverse effects of supply disruptions can be based on the structural analysis of supply chain composition, i.e. the degree of its geographical (spatial) concentration and the diversification of suppliers (number of countries/firms). The risk is proportional to the degree of geographic concentration of supplying countries and firms (Schwellnus et al., 2023). The electric mobility sector, encompassing the production of electric batteries, charging stations, electric buses for public transportation or BEVs, is prone to supply chain disruption because of the high geographic concentration of component manufacturing and critical minerals mining in the electric power supply chain. For instance, over one-third (35%) of exports of electric batteries (the world's 31st most traded product out of 1217) come from just one country: China.[2] The distribution of critical minerals' supply, vital in BEV chain, is also extremely concentrated: for instance, 45% of global lithium supply comes from Australia (followed by Chile—28% and China—10%).[3]

With President Joe Biden's new import duties imposed on various Chinese goods in May, the EU is presented with a hard choice. Accelerating the transition to electric mobility based on relatively cheap imports of Chinese EVs risks undermining the EU manufacturing base and raises security concerns. Large-scale opening of the EU's internal market for EVs produced in economies with lower environmental and labour standards tilts the global playing field and disadvantages of European producers. It also impedes EU's domestic abilities to expand capacities in key areas like semiconductors and clean tech. Chinese EVs are competing with European ones due to much lower prices. After a 15% price cut, BYD's flagship SUV costs €41,000 in Germany. Its closest competitors in the segment cost approx. EUR 4000 more expensive. The European Commission announced in March 2024 that it has sufficient evidence that Chinese electric vehicles are being subsidized by the authorities, which gives them the upper hand at the EU market.

Local Ruptures in Electric Mobility Development

The PwC Strategy & Readiness Index is a synthetic indicator that measures the level of maturity of a country in relation to the transition to electric cars. The four dimensions taken into consideration are: government incentives; charging infrastructure; the supply of electric vehicles; and consumer demand. Poland ranks last, mainly due to the lack of public charging infrastructure. France has an intermediate position far behind Norway and Switzerland, particularly due to the weakness of the second-hand market.[4] Thus, even in two countries, such as Poland and France, with high/low GDP per capita and different extents of renewable energy and infrastructure development, the cost is still perceived as high. This conclusion came also out of the analyses within the ITEM project in the partner city of Poznań in Poland (ITEM 2020–2024).[5] The majority of the interviewed residents declared no financial capacity to buy an EV. Experts taking part in a workshop carried out in Poznań (ITEM 2020–2024) reaffirmed this observation. Not only in Poland but also in France, the cost of a new private EV is perceived as very high.

With governmental subsidies for private BEV purchase, a concern about social justice implications ensue questioning public grants for luxury goods, such as BEVs. Similarly, at the urban level, investments in electric or hydrogen buses come at a high cost, and, even if subsidised, often lead to the increase in public transport fares. Charging infrastructure is another area where high investment costs are difficult to avoid, if the effects of scale are to be achieved, and become problematized by the city administration and different stakeholders (ITEM 2020–2024).

The cost of the development of the needed infrastructure is also high. In France, to connect new solar and wind farms and charging stations for electric cars, as well as improve cable resistance to extreme weather events, French electricity grid operator Enedis plans to invest more than 5 billion euros a year by 2032, compared with less than 3 billion over the past forty years.[6] This promises to inflate the costs of the overall system. The management of high and extra-high voltage lines, Réseau de Transport Electrique (RTE, Electrical Transmission Network), will also come at a huge operating cost due to the need to connect power lines to the grid.[7] Locally, a vision of electric mobility transition may not be attractive because of its costs, the electrical network's unsuitability in rural areas but also because of a rising awareness of its ecological impact globally.

Conclusion and Recommendations

Building on our experience of years-long engagement in collaborative and interdisciplinary research projects, we zoomed in on three key threats: the lack of **coherence** of the undertaken policies, potential **ruptures** across regions and scales, and overlooked **multiplicity** of factors and contexts that determine the transition to electric mobility. To recapitulate key findings, we have grouped them in a Table 4.1, together with proposed recommendations and addressed the audience to break out of silo logic of policymaking and stimulate cross-sectoral exchange. To close the gap between public policymakers and a vast swath of public entrepreneurs and problem solvers working in the field of EM, we suggested the particular audience groups which might be most interested in the exact proposals. Some of our suggestions, like fostering greater transparency of the origins of the product and scrutinize the whole supply chain, have been partly addressed in the very moment of our joint work, with the introduction of the Corporate Sustainability Due Diligence Directive (CSDDD) in the EU. Others, like strengthening the resilience of global value chains by analysing geopolitical risks and environmental impacts in non-European regions, still wait for a more decisive action.

Table 4.1 Identified problems, recommendations to solve it, and the addressed audiences

Identified problem	Recommendations to solve it	Addressed audience other than EU and national policymakers
Ethical questions of ICEVs and BEVs usage	• Standardise LCA and laboratory tests • Improve transparency of product origins, including components and raw materials used in manufacturing	• Data providers, researchers • Producers and retailers • Opinion formers
Making supply chains more resilient to economic and geopolitical impacts	• Increase investment in R&D to maximise recycling and the use of alternative raw materials • Establish a common framework for used cars recycling • Increase investment in R&D in the second-life of batteries • Include EVs into the right-to-repair rule • Conduct supply chain analysis for key resources for electric mobility transition to trace the most vulnerable nodes in the supply system and address the economic feasibility of transition • Focus on the resilience of global value chains by analysing geopolitical risks and environmental impacts in non-European regions • Establish a Geopolitical Risk Body within the EC to scrutinise geopolitical threats to electric mobility supply chains	• Financial institutions • Researchers • Producers • MPs

Identified problem	Recommendations to solve it	Addressed audience other than EU and national policymakers
Local and regional ruptures in electric mobility development	• Create local and regional institutions to monitor, communicate, and negotiate implications of electric mobility transition for local economic value and social justice to feed in policy processes at regional and national levels • Apply EU level Just Transition instruments to electric mobility transition in European regions	• Experts • Local administration • Local stakeholders

Notes

1. IMUMNC: Improved urban mobility towards climate neutrality under new working habits and transport modes, ERA-NET Urban Accessibility and Connectivity and the National Science Center of Poland, grant number 2022/04/Y/HS4/00135; ITEM: Inclusive Transition towards Electric Mobility, ERA-NET Urban Accessibility and Connectivity and the National Science Center of Poland, grant number 2020/02/Y/HS4/00078; Powering the world: STS and anthropology towards social studies on new energies, National Science Center, OPUS, grant number 2017/25/B/HS6/00880; Zeitenwende. Change and its contestation in the model of strategic foreign economic policy of Germany, National Science Center, SONATA, grant number 2022/47/D/HS5/03380; RETHINK-GSC: Rethinking Global Supply Chains: measurement, impact and policy EU's Horizon Europe, grant number 101061123. Tech-Spec: Technological specialisation and productivity divergence in the age of digitalisation, automation and AI, National Science Center, OPUS, grant number DEC-2020/37/B/HS4/01302; Social Sciences and Humanities for Advancing Policy in European Energy, European Commission, Horizon 2020, grant number 731264; Energy Social sciences & Humanities Innovation Forum Targeting the SET-Plan, European Commission, Horizon 2020, grant number 826025.
2. Data from The Observatory of Economic Complexity, statistics refer to 2021. Source: https://oec.world/en/profile/hs/electric-batteries [date of access: 15 January 2024].
3. Data from Statista, statistics refer to 2020. Source: https://www.statista.com/statistics/968980/lithium-supply-distribution-worldwide-by-country/ [date of access: 15 January 2024]. Primary source: Deutsche Bank Research.
4. https://www.pwc.fr/fr/espace-presse/communiques-de-presse/2023/octobre/l-electrique-depasse-le-diesel.html.
5. https://researchcentre.amu.edu.pl/project/item/ or https://www.itemresearch.org/.
6. https://www.enedis.fr/new-electric-france-2027-and-2032-enedis-publishes-preliminary-document-its-future-network.

7. https://assets.rte-france.com/prod/public/2023-02/2023-01-19-webinaire-rte-nos-perspectives-investissement-reseau-support-de-presentation.pdf.

REFERENCES

Alyavina, E., Nikitas, A., & Tchouamou Njoya, E. (2020). Mobility as a service and sustainable travel behaviour: A thematic analysis study. *Transportation Research Part F: Traffic Psychology and Behaviour, 73*, 362–381. https://doi.org/10.1016/j.trf.2020.07.004

Antràs, P., & Chor, D. (2022). Global value chains. *Handbook of International Economics, 5*, 297–376. https://doi.org/10.1016/bs.hesint.2022.02.005

Baldwin, R., & Freeman, R. (2022). Risks and global supply chains: What we know and what we need to know. *Annual Review of Economics, 14*(1), 153–180.

Banerjee, A., & Duflo, E. (2019). *Good economics for hard times. Better answers to our biggest problems*. Allen Lane.

Bolet, D., Green, F., & González-Eguino, M. (2023). How to get coal country to vote for climate policy: The effect of a "Just Transition Agreement" on Spanish election results. *American Political Science Review*. Published online 2023, 1–16. https://doi.org/10.1017/S0003055423001235

Bolitho, A. (2023). *Europe in race to secure raw materials critical for energy transition*. Euronews. https://www.euronews.com/business/2023/03/07/europe-in-race-to-secure-raw-materials-critical-for-energy-transition

Bradford, A. (2020). *The Brussels effect: How the European Union rules the world*. Oxford University Press.

Dorn, F. E., Hafner, R., & Plank, C. (2022). Towards a climate change consensus: How mining and legitimize green extractivism in Argentina. *The Extractive Industries and Society, 11*, 101130. https://doi.org/10.1016/j.exis.2022.101130

Enedis. (2023). *New Electric France 2027 and 2032: Enedis publishes the preliminary document to its future Network Development Plan for electricity distribution*. https://www.enedis.fr/new-electric-france-2027-and-2032-enedis-publishes-preliminary-document-its-future-network

European Commission. (2020). *Study in the EU's list of Critical Raw Materials—Final Report*. https://doi.org/10.2873/904613

European Commission. (2023). *A new outlook on the climate and security nexus: Addressing the impact of climate change and environmental degradation on peace, security and defence*. Joint Communication to the European Parliament and the Council. https://www.eeas.europa.eu/eeas/joint-communication-climate-security-nexus_en

Flipo, A., Sallustio, M., & Ortar, N. (2023). Can the transition to sustainable mobility be fair in rural areas? A stakeholder approach to mobility justice. *Transport Policy, 139*. https://doi.org/10.1016/j.tranpol.2023.06.006

Goedkoop, M., Heijungs, R., Huijbregts, M., Schryver, A., Struijs, J., & Zelm, R. (2008). *ReCiPE 2008 : A life cycle impact assessment method which comprises harmonised category indicators at the midpoint and the endpoint level.*

gridX. (2023). *EV Charging Report 2023.* https://www.gridx.ai/resources/ev-charging-infrastructure-report-europe-2023

Henderson, J. (2020). EVs are not the answer: A mobility justice critique of electric vehicle transitions. *Annals of the American Association of Geographers, 110*(6), 1993–2010. https://doi.org/10.1080/24694452.2020.1744422

Item. (2023). *Inclusive Transition towards Electric Mobility (ITEM).* https://www.itemresearch.org/

Item. (2024). *Inclusive Transition towards Electric Mobility (ITEM).* https://researchcentre.amu.edu.pl/project/item/

Jabko, N. (2006). *Playing the Market: A Political Strategy for Uniting Europe, 1985–2005.* Cornell University Press.

Lèbre, É., Stringer, M., Svobodova, K., Owen, J. R., Kemp, D., Côte, C., Arratia-Solar, A., & Valenta, R. K. (2020). The social and environmental complexities of extracting energy transition metals. *Nature Communications, 11*, 4823. https://doi.org/10.1038/s41467-020-18661-9

Meckling, J., & Allan, B. B. (2020). The evolution of ideas in global climate policy. *Nature Clinical Practice Endocrinology & Metabolism, 10*, 434–438. https://doi.org/10.1038/s41558-020-0739-7

Observatory of Economic Complexity. (2021). *Electric Batteries.* https://oec.world/en/profile/hs/electric-batteries

Ortar, N., & Ryghaug, M. (2019). Should all cars be electric by 2025? The electric car debate in Europe. *Sustainability, 11*(7), 1868. https://doi.org/10.3390/su11071868

Schwellnus, C., Harambourne, A., Samek, L., Chiapin Pechansky, R., & Cadestin, C. (2023). Global value chain dependencies under the magnifying glass. *OECD Science, Technology and Industry Policy Papers.* https://doi.org/10.1787/23074957

Sovacool, B. K., Ali, S. H., Bazilian, M., Radley, B., Nemery, B., Okatz, J., & Mulvaney, D. (2020). Sustainable minerals and metals for a low-carbon future. *Science, 367*(6473), 30–33. https://doi.org/10.1126/science.aaz6003

Statista. (2020). https://www.statista.com/statistics/968980/lithium-supply-distribution-worldwide-by-country/

Tagliaferri, C., Evangelisti, S., Acconcia, F., Domenech, T., Ekins, P., Barletta, D., & Lettieri, P. (2016). Life cycle assessment of future electric and hybrid vehicles: A cradle-to-grave systems engineering approach. *Chemical Engineering Research and Design*, *112*(Supplement C), 298–309. https://doi.org/10.1016/j.cherd.2016.07.003

VDI. (2023). *VDI-Analyse der CO2 äq-Emissionen von Pkw mit verschiedenen Antriebssystemen* (Blaue Papiere, 19, VDI e.V. Düsseldorf). https://doi.org/10.51202/9783949971747

Open Access This chapter is licensed under the terms of the Creative Commons Attribution 4.0 International License (http://creativecommons.org/licenses/by/4.0/), which permits use, sharing, adaptation, distribution and reproduction in any medium or format, as long as you give appropriate credit to the original author(s) and the source, provide a link to the Creative Commons license and indicate if changes were made.

The images or other third party material in this chapter are included in the chapter's Creative Commons license, unless indicated otherwise in a credit line to the material. If material is not included in the chapter's Creative Commons license and your intended use is not permitted by statutory regulation or exceeds the permitted use, you will need to obtain permission directly from the copyright holder.

PART III

The Potential of the Transformation of Transport Services and Vehicle Technologies to Contribute to the Transition

CHAPTER 5

Promoting Sustainable Urban Mobility Through Implementation of Electric Buses: A Case Study of Ostrava

Marek Krumnikl, Adam Červenka, Filip Lapuník, and Luboš Mikula

Abstract We recommend promoting sustainable urban mobility through the implementation of electric buses. To achieve this policy recommendation, we propose to take into account the following: (1) Cities should prioritise the transition to electric and CNG (compressed natural gas) buses for sustainable public transport, considering both ecological and

---

M. Krumnikl (✉) · A. Červenka
University of Ostrava, Ostrava, Czech Republic
e-mail: marek.krumnikl@osu.cz

A. Červenka
e-mail: adam.cervenka@osu.cz

F. Lapuník · L. Mikula
VSB-Technical University of Ostrava, Ostrava, Czech Republic
e-mail: filip.lapunik.st@vsb.cz

L. Mikula
e-mail: mik0345@vsb.cz

economic impacts; (2) Cities undergoing transition due to the decline of heavy industry can benefit from subsidies supporting sustainable mobility, thereby modernising their vehicle fleet, which has positive ecological and economic impacts; (3) Embrace a multilevel governance approach learning from Ostrava's experience and utilise tailored regional strategies supported by national and EU-level initiatives; (4) Regions with cleaner electricity generation (300 to 600 gCO_2eq/kWh) can benefit from using electric buses more than other regions and see a pronounced effect on emissions and sustainability; and (5) As for social impact, passengers in public transport see comfort as the biggest priority, not the environmental impact.

Keywords Sustainable urban mobility · Electric buses · Ecologisation

Introduction

In the 1960s, car ownership in Europe increased, shaping urban development around automobiles. Recently, there's been a shift towards sustainable mobility, prioritising environmental sustainability, efficiency, and inclusivity (Curtis, 2020). European and Czech policies now emphasise low-emission transport. Cities are focusing on modernising and ecologising public transport. This study examines the potential electrification of Ostrava's bus fleet, considering emissions, energy consumption, and economic factors. Ostrava was selected due to its varied public transportation system and its similarities with other European cities from regions impacted by structural changes, which had a history of heavy industry throughout the twentieth century. This industrial legacy continues to influence these regions to this day.

The main goal of the research is to evaluate the efficiency of compressed natural gases (CNG) and electric buses from an economic and ecological point of view. The other goal is to find out how successful is Ostrava in fulfilling goals established in strategic documents and the impact of ecologisation of vehicle fleets to the city transport company (DPO) and the city itself. SSH and STEM researchers collaborated on a single case study of the city Ostrava to assess the use of low-emission buses and explore its potential as a model for similar cities. They conducted eight interviews with members of the city council and

Public Transport Company stakeholders, exploring how Ostrava fulfils the objectives outlined in strategic sustainability documents. This exploration also covered public perceptions of fleet modernisation and the ecological and economic benefits of the new bus fleet. SSH researchers provided theoretical frameworks and performed a discursive analysis based on these documents and the conducted interviews, while STEM researchers analysed the efficiency and impact of electric and CNG buses compared to diesel. This quantitative analysis of secondary data was conducted using information gathered from local authorities, transport companies and previously published research papers which focus on similar topics to assess the impact utilisation of alternative propulsion systems may have on public transport, especially the CO_2 emissions in regard to the energy consumption and emissions created during electricity generation processes. Together SSH and STEM researchers evaluated the ecological, economic, and social benefits of the fleet's modernisation, aligning with EU, national, and regional strategies.

The study focuses on Ostrava, the third-largest Czech city with around 280,000 inhabitants in 2021 (ČSÚ, 2021), and part of the country's second-largest agglomeration. Ostrava was selected as a case study because it is a medium-sized city in Central Europe with a historical background of heavy industry, featuring all modes of transportation. This makes Ostrava a representative example for many European cities facing similar challenges.

Understanding Ostrava requires insights into its historical and socio-geographical evolution. Formed from 34 municipalities and significantly developed during the industrial revolution through coal mining and metallurgy (Bakala et al., 1993), Ostrava's urban design diverges from traditional Czech cities, resulting in a polycentric structure and it is considered as structurally affected region (Kuta & Endel, 2015). Despite this, Ostrava exerts strong regional influence, with a daily influx of over 20,000 people for work or study from nearby cities (ČSÚ, 2021).

Ostrava's public transport system, part of the ODIS (regional transport network), includes 68 lines of trams, trolleybuses, and buses, covering 914.7 km (DPO a.s., 2022). The extensive network underscores Ostrava's regional importance. The Transport Company of Ostrava has modernised its fleet, reducing diesel buses by 85% between 2014 and 2022, with 227 CNG and 24 electric buses. The average bus age in 2022 was 5.9 years (DPO a.s., 2015, 2022).

Sustainable Transportation in the Context of EU Policy and Local Implementation: The Case of Ostrava

In advancing sustainable urban mobility, mere reliance on technological innovation is insufficient. The complex interplay of strategy creation and policy implementation, as highlighted by (Von Schönfeld & Ferreira, 2021), necessitates a holistic approach that addresses the diverse needs of citizen groups (Berger et al., 2014). However key players such as the European Commission, in their 2011 White Paper, advocate for a paradigm shift, stating, "A different trajectory is essential for transport's future development over the next 40 years" (European Commission, 2011). This paradigm shift, integral to the Europe 2030 Strategy, cascades from EU-level directives to regional and urban policymaking. The European Commission's Mobility Strategy articulates that the European Union is committed to ensuring that, by the year 2030, all scheduled collective travel for distances under 500 kilometres will achieve carbon neutrality. Furthermore, the strategy delineates that by the year 2050, it is projected that nearly all automobiles, vans, buses, and new heavy-duty vehicles within the EU will be zero emission vehicles. Current data indicate that transportation within the European Union is responsible for producing approximately 25% of greenhouse gas emissions. The overarching aim of the European Green Deal is to effectuate a reduction in these emissions by 90% by the year 2050, thereby substantially mitigating the transportation sector's environmental impact and contributing to the EU's ambitious climate objectives (commission.europa.eu, 2024; transport.ec.europa.eu, 2024).

Key national documents in Czechia, such as the Czech Republic 2030, (Sustainable Urban Mobility Plan) SUMP 2.0, White Book 2015, and The Transport Policy of the Czech Republic 2021–2027, reflect a commitment to sustainable mobility. These policies, particularly the SUMP 2.0, advocate a lifestyle-centric approach, intertwining health and transportation. If these strategies are applied, in urban contexts like Ostrava, public transport could account for up to 45% of total passenger transportation share, significantly influencing city dynamics.

The National Action Plan for Clean Mobility aligns with the EU Directive (2014/94/EU), promoting electromobility, CNG, and hydrogen fuels. This plan underscores a phased technology adoption, prioritising electromobility and CNG, followed by LNG (liquefied natural gases) and hydrogen. Cities are tasked with aligning their transport services with SUMP goals, including the adoption of low-emission measures and utilisation of subsidy programs. Ostrava exemplifies this approach, successfully securing grants such as the Integrated Regional Operational Programme (IROP) to facilitate fleet modernisation with alternative fuels (akademiem estskemobility.cz, 2023).

The regional strategy from the Moravian-Silesian Region, particularly the Strategy for the Development of the Moravian-Silesian Region 2019–2027, focuses on sustainable transport and CO_2 emission reduction. Ostrava's alignment with this strategy is evident in its commitment to reducing bus emissions. The city's strategic plans aim for 95% emission-free or low-emission public transport vehicles by 2025 and complete diesel phase-out by the end of 2020s. This aligns with Ostrava's SUMP, indicating a well-coordinated approach between local and regional policy goals.

The rapid modernisation and greening of Ostrava's public transport, reduction of its diesel fleet by more than 85% between 2014 and 2022, demonstrate the city's commitment to ecological improvement and a cleaner image. Today, Ostrava's public transport relies on alternative power, with 29 electric and 225 CNG buses, addressing nearly half of the city's public transport needs. Despite the significant technological advancements in transitioning bus fleets from diesel to alternative energy sources, this shift may not result in perceptible improvements in terms of passenger comfort. According to the results of interviews, it is evident that passenger comfort during travel remains a priority over the environmental attributes of the transportation medium. Consequently, the type of energy powering the journey is often of minimal concern to passengers, who prioritise comfort above all.

Energy Consumption and Emission Analysis

In 2022, DPO vehicles carried 85,926 passengers (Ostrava, 2022). Despite buses having 10,000 fewer passengers than trams and trolleybuses, they remain vital for connecting city outskirts and suburban areas. Buses lead in total line length, number of services, and fleet size. Economically, the average cost per km for buses was 64.35 Kč, 37% cheaper than trams.

This study compares Solaris 12 CNG and Solaris 12 electric buses (these models were chosen based on the fact that they are very similar in dimensions and used in the DPO fleet). According to the interview both types operate all-day shifts with rapid charging for electric buses. Interviewees reveal passengers do not notice significant differences in comfort or quality, despite the shift to a sustainable bus network. Key factors for comparison include environmental impact and measurable advantages or disadvantages of each propulsion system.

There are currently three locations with rapid charging in Ostrava, located in strategic locations with direct connection to tram lines. The most advanced rapid charging hub is located 1.5 km from the city centre on Valcharska Street. Hub is capable of charging up to 15 battery electric buses (BEB) per hour. Chargers comply with the OppCharge and ISO/IEC 15,118, to preserve maximum compatibility in the future. Pantograph chargers are used, bringing time-saving and user comfort. With 300 kW installed peak power in the location, system is designed for 5–10 minute long charging intervals. End station charging enables the producer to use smaller in-vehicle batteries, bringing more passenger space, faster charging, and lower purchase prices, together with high infrastructure utilisation during the day. One of the great disadvantages is the need for charging infrastructure and the higher demand for vehicle organisation and planning. (Gouiaa, 2018).

Ostrava's vehicle depots have two CNG fill-up stations, with the Hranečník garage facility serving four vehicles at once with an 8-minute fill-up time. The total cost of the CNG filling station was 54 million CZK in 2015 (80 million CZK in 2023), while the charging station cost 21.5 million CZK in 2023 (DPO a.s., 2017, 2021). The vehicle cost for a 12 m CNG bus was 6.97 million CZK in 2017 (10 million CZK in 2023), and 10.9 million CZK for a 12 m electric bus in 2023, making the purchase cost differences insignificant. (DPO a.s., 2021).

Vehicle operational costs depend mostly on fuel costs. With varying spot CNG prices, there is high volatility in operating costs. With an average fuel consumption of 31 kg/100 km for a 12 m CNG bus, the cost can vary from 86.8 CZK–372 CZK per 100 driven kilometres. Determination of final BEB fuel costs is difficult mainly due to the unknown charging efficiency. According to DPO, the average cost per 100 driven kilometres is 100 CZK, for an average energy price of 2500 CZK per 1MWh. The real BEB power consumption should be in the range between 1.5 and 3 kWh/km.

A method for direct comparison of emissions produced by diesel, CNG, and LNG buses already exists (MMR ČR, 2017). This method can only be used to find out the emissions saved on a local level but cannot provide enough insight and advice on the relevance and sustainability benefits of electrification of buses. A way of showcasing the impact of this action and to put it into the European perspective of transportation, energy, and power sector is to compare both options (CNG and electric buses) on similar terms and not just think about the emissions saved by using electric buses (thinking of this system as a zero emission).

CO_2 produced is an interesting factor since it universally showcases the direct impact of switching to BEB and can serve for a scenario analysis based on external factors. The annual average amount of CO_2 equivalent produced per kWh in the EU in 2023 differed by a large margin (two countries that serve as an example are: France—47 gCO_2/kWh, Poland—800 g CO_2/kWh) (Electricity Maps, 2024). In Czechia in 2023 the annual average was 540 grams of CO_2eq/kWh consumed (Electricity Maps, 2024).

Another study indicates that CNG buses can generate approximately 970 to 1300 grams of CO_2 per kilometre driven, based on experiments conducted in real-life conditions. (Prati et al., 2022) In comparison, decommissioned diesel buses emit between 1300 to 1600 grams of CO_2 per kilometre (Rosero et al., 2020). Therefore, even at the upper limit of emissions, CNG buses produce less CO_2 than their diesel counterparts.

Based on this data we estimated that in Ostrava the electric buses produce about the same amount of CO_2/km as the CNG alternative in case they have the energy consumption of 1.7 to 2.4 kWh/km. If the consumption is lower the usage of electric buses is beneficial.

To put the scale of public transport into perspective and to allow for a potential comparison of Ostrava to other cities, the average vehicle cumulative kilometres travelled annually by public transport buses in the past 3 years (2020–2022) was 16,805 thousand kilometres. The average turnover speed in the past 3 years (2020–2022) is 18.4 km/h (DPO a.s., 2022; Krumnikl et al., 2024).

The following Fig. 5.1 showcases how the amount of CO_2eq/kWh of electricity consumed may affect the amount of CO_2 produced per km. This illustrates this issue on a global scale and how this could affect other countries with a different value of CO_2eq/kWh consumed.

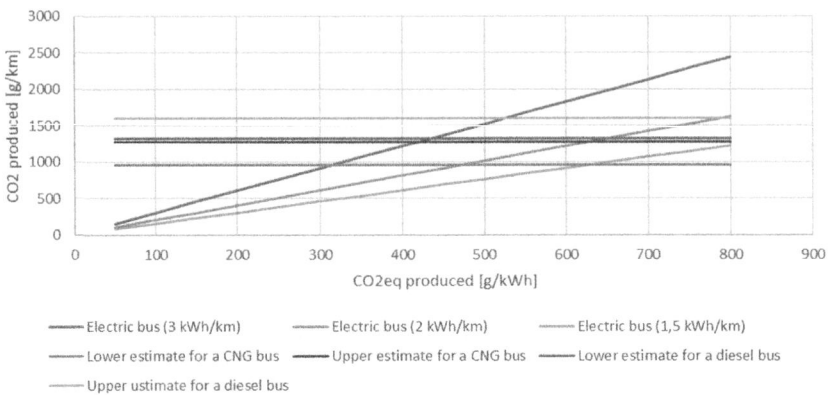

Fig. 5.1 Grams of CO_2/km produced based on energy consumption of the vehicle (Color figure online)

Leveraging the Ostrava Model for EU-Wide Sustainable Mobility Policies

Below is a list of recommendations for various governance levels:

1. European Commission: Use the Ostrava model as a benchmark in EU mobility and transport policy frameworks. For example, Ostrava's successful integration of the Integrated Regional Operational Programme (IROP) funding to upgrade its public transport fleet with CNG and electric buses can serve as a model for other member states and regions. This approach supports the EU's goal of a 90% reduction in transport emissions by 2050.
2. National Governments: Align national strategies with EU sustainability goals, as demonstrated by the Czech Republic's integration of SUMP 2.0 with the Europe 2030 Strategy. Develop frameworks and financial mechanisms for cities transitioning to clean mobility, similar to the subsidies provided for Ostrava's public transport overhaul.
3. Regional Authorities: Follow the Moravian-Silesian Region's example of supporting city-level initiatives, which focus on sustainable transport and CO_2 emission reduction, offering guidance and resources to cities like Ostrava for implementing green transport measures.

4. City Administrations: SUMPs that resonate with regional and EU policies.
5. Cities should look at Ostrava's approach to reducing its diesel fleet and increasing low-emission vehicles, significantly contributing to the regional goal of a cleaner environment.

By embracing this multilevel governance approach, the EU and its member states can learn from Ostrava's experience to effectively move towards sustainable urban mobility and meet carbon emission reduction targets. Ostrava's success story serves as an example that showcases how tailored regional strategies, coupled with national and EU-level support, can lead to significant improvements in urban transport sustainability.

Conclusion

Initial investments in vehicle fleets and infrastructure are substantial, and without subsidy support, they are often unfeasible for many cities within the EU context. However, when cities actively utilise a range of subsidy programs and succeed in modernising and transitioning to sustainable energy sources, it becomes evident that, in the long term, this approach is more cost-effective than traditional propulsion systems, yielding significant savings.

The exploration of Ostrava's transition to electric and CNG buses reveals a complex interplay between technology, policy, and environmental considerations. Our findings suggest that the benefits of electric buses are significantly influenced by the carbon intensity of the power supply. In regions where electricity generation is cleaner (emitting between 300 to 600 gCO_2eq/kWh may be considered as a good benchmark), electric buses present a more sustainable option. However, in areas with higher carbon intensity in electricity generation, the ecological advantages may not be as pronounced. The benchmark for BEB being the more sustainable option to diesel powered buses is around 400–800 gCO_2eq/kWh.

The case study of Ostrava exemplifies the potential and challenges of urban transit transformation in a post-industrial city context. The city's success in fleet modernisation, driven by strategic policy alignment and external funding, highlights the importance of comprehensive planning and support at multiple governance levels. The economic aspect, particularly the comparison of operational costs between CNG and electric buses, underscores the need for a long-term perspective when evaluating the feasibility of transitioning to sustainable transport options.

This study highlights that the suitability of electric buses varies across Europe due to differences in CO_2 emissions per kWh of electricity. Cities must assess their unique energy landscapes when considering electrification.

The transition to electric buses in Ostrava shows potential for reducing urban emissions, but it is not universally applicable. Successful implementation requires understanding local energy systems, policy support, and considering long-term economic and environmental impacts. This study guides other cities in sustainable transport transitions, emphasising critical factors for success. For cities like Ostrava, subsidies can help manage the initial investment in vehicles and infrastructure, making sustainable transport solutions more cost-effective in the long term. This insight is particularly relevant for EU cities, where upfront costs can be a barrier without subsidies. Thus, this case study offers a model for similar urban centres aiming for sustainable public transportation.

Currently, the utilisation of modern technologies associated with the electrification of buses in Ostrava is aimed locally. From interviews with representatives of local public transport companies, it became apparent that purchasing buses and the corresponding infrastructure collectively is advantageous to avoid compatibility issues between the buses and infrastructure. Given that these acquisitions are currently primarily regulated by the Public Transport Company (DPO) within the city's jurisdiction, it poses a significant challenge for infrastructure sharing with other operators active in Ostrava. Another factor is the high utilisation of the existing infrastructure by DPO vehicles. Should the procurement of new vehicles and expansion of infrastructure be addressed at a regional or national level, it could facilitate better infrastructure sharing among various operators. However, it is also crucial to plan the locations of charging stations and strategically schedule the charging times for different bus routes to optimise usage and ensure efficiency.

References

AF-CITYPLAN. (2014). *Integrovaný plán mobility Ostrava část I.-strategická*. Retrieved January 15, 2024, from https://mobilita-ostrava.cz/finalni-verze-navrhove-casti-dokumentu-integrovany-plan-mobility-ostrava/
akademiemestskemobility.cz. (2023). Akademiemestskemobility.cz. *Akademie Městské Mobility*. https://www.akademiemobility.cz/dokumenty
Bakala, J., Jiřík, K., Jiřík, K., Gracová, B., Klíma, B., & Borák, M. (Eds.). (1993). *Dějiny Ostravy* (1st ed.). Sfinga.
Berger, G., Feindt, P. H., Holden, E., & Rubik, F. (2014). Sustainable mobility—Challenges for a complex transition. *Journal of Environmental Policy & Planning, 16*(3), 303–320. https://doi.org/10.1080/1523908X.2014.954077
commission.europa.eu. (2024). *Providing efficient, safe and environmentally friendly transport*. Retrieved February 30, 2024, from https://commission.europa.eu/strategy-and-policy/priorities-2019-2024/european-green-deal/transport-and-green-deal_en
ČSÚ. (2021). *Czso.cz*. https://www.czso.cz/
Curtis. (2020). *Handbook of sustainable transport* (1st ed.). Elgar.
Dopravní podnik Ostrava a.s. (2015). *Výroční zpráva 2015*. Ostrava. Retrieved from https://www.dpo.cz/soubory/spolecnost/v-zpravy/2015.pdf
Dopravní podnik Ostrava a.s. (2017). *Koupě 24 ks jednočlánkových autobusů na CNG pohon*. Retrieved January 10, 2024, from https://smlouvy.gov.cz/smlouva/12987240
Dopravní podnik Ostrava a.s. (2022). *Výroční zpráva 2022*. Ostrava. Retrieved from https://www.dpo.cz/soubory/spolecnost/v-zpravy/2022.pdf
Dopravní podnik Ostrava a.s. (2021, February 19). *Koupě 24 ks jednočlánkových elektrobusů a 2 ks nabíjecích stanic*. Retrieved January 10, 2024, from https://smlouvy.gov.cz/smlouva/15898239
Electricity Maps. (2024, January 28). *Electricity Maps. Electricity Maps | Reduce carbon emissions with actionable electricity data*. Retrieved January 28, 2024, from https://www.electricitymaps.com/
European Commission. (2011). *Roadmap to a single European transport area – Towards a competitive and resource efficient transport system*. European Union. https://eur-lex.europa.eu/legal-content/EN/TXT/?uri=CELEX%3A52011DC0144
Gouiaa, A. (2018, December 8). *City-scale, agent-based modelling & analysis of an electric public bus transport system*. https://doi.org/10.13140/RG.2.2.12633.16482

Krumnikl, M., Červenka, A., Lapunik, F., & Luboš, M. (2024). Promote sustainable urban mobility through implementation of electric bus: A case study of Ostrava. *Zenodo*. https://doi.org/10.5281/zenodo.11401167

Kuta, V., & Endel, S. (2015). *Ostrava jako regionální metropole* (1st ed.). Statutární město Ostrava.

Ministerstvo pro místní rozvoj ČR. (2017, November 3). *Metodické listy indikátorů_SC1.2*.

Ostrava. (2017). *Strategický plán rozvoje statutárního města Ostravy 2017–2023*. Retrieved January 20, 2024, from https://fajnova.cz/strategicky-plan-2017-2023/

Ostrava. (2022). *Informace o dopravě v Ostravě 2022*. Retrieved January 20, 2024, from https://www.ostrava.cz/cs/urad/magistrat/odbory-magistratu/odbor-dopravy/oddeleni-silnic-mostu-rozvoje-a-organizace-dopravy/informace-o-doprave/Sbornkinformacodoprav2022.pdf

Prati, M. V., Costagliola, M. A., Unich, A., & Mariani, A. (2022, December). *Emission factors and fuel consumption of CNG buses in real driving conditions. Transportation Research Part D: Transport and Environment*, 113. ScienceDirect. https://doi.org/10.1016/j.trd.2022.103534

Rosero, F., Fonseca, N., López, J.-M., & Casanova, J. (2020). Real-world fuel efficiency and emissions from an urban diesel bus engine under transient operating conditions. *Applied Energy, 261*, 114442. Retrieved May 27, 2024, from https://www.worldtransitresearch.info/research/7810/

transport.ec.europa.eu. (2024). *Mobility Strategy*. Retrieved February 30, 2024, from https://transport.ec.europa.eu/transport-themes/mobility-strategy_en

Von Schönfeld, K. C., & Ferreira, A. (2021). Urban planning and European innovation policy: Achieving sustainability, social inclusion, and economic growth? *Sustainability, 13*(3), 1137. https://doi.org/10.3390/su13031137

Open Access This chapter is licensed under the terms of the Creative Commons Attribution 4.0 International License (http://creativecommons.org/licenses/by/4.0/), which permits use, sharing, adaptation, distribution and reproduction in any medium or format, as long as you give appropriate credit to the original author(s) and the source, provide a link to the Creative Commons license and indicate if changes were made.

The images or other third party material in this chapter are included in the chapter's Creative Commons license, unless indicated otherwise in a credit line to the material. If material is not included in the chapter's Creative Commons license and your intended use is not permitted by statutory regulation or exceeds the permitted use, you will need to obtain permission directly from the copyright holder.

CHAPTER 6

Improving Rural Quality of Life by Combining Public Transportation with Demand Responsive Transport Systems

József Pál Lieszkovszky, Dániel Tordai, Daniel Hörcher, Tamás Fleischer, and András Munkácsy

Abstract We recommend improving rural quality of life by combining public transport with demand responsive transport systems. To achieve this policy recommendation, we propose the following: (1) Promote the creation of demand responsive transport (DRT) systems that fit well into the region's or nation's larger transport strategy and system, are

J. P. Lieszkovszky (✉) · D. Tordai · A. Munkácsy
KTI Hungarian Institute for Transport Sciences and Logistics, Budapest, Hungary
e-mail: lieszkovszky.jozsef@kti.hu

D. Tordai
e-mail: tordai.daniel@kti.hu

A. Munkácsy
e-mail: munkacsy.andras@kti.hu

J. P. Lieszkovszky
Corvinus University of Budapest, Budapest, Hungary

© The Author(s) 2024
I. Keseru et al. (eds.), *Strengthening European Mobility Policy*,
https://doi.org/10.1007/978-3-031-67936-0_6

cost effective and sustainable in the long run; (2) Identify and clarify the societal goals that DRT operations are intended to support. Subsidising DRT operations cannot be justified just for their own sake; (3) Define the intended demand intensity to be served with this transport service explicitly; (4) Integrate DRT services with complementary incentive mechanisms that preserve the benefits of the spatial concentration of the population and workplaces; (5) Ensure a reasonable level of long-term public funding for DRT systems right from the planning phase; and (6) Reduce the barriers of entry to the DRT market and develop a service model, enabling the utilisation of excess transport capacity that local SMEs and public institutions as minibus owners may have.

Keywords Demand responsive transport · Rural mobility · Sustainable urban mobility plan (SUMP)

Introduction

In this chapter, we examine demand responsive transport (DRT) schemes and provide policy recommendations on supporting DRT solutions in rural areas to achieve better outcomes.

DRT is defined as "an intermediate form of public transport, somewhere between a regular service route that uses small low floor buses and variably routed, highly personalised transport services offered by taxis" (Brake et al., 2004: 324). The promise of DRT is to revitalise public transport in sparsely populated areas by offering denser supply and increasing flexibility. We recognise the common assertion that it can improve service quality in regions rarely served by traditional public transport, provide service in currently unserved areas at similar service levels as in regions

---

D. Hörcher
Imperial College London, London, UK
e-mail: d.horcher@imperial.ac.uk

T. Fleischer
Fleischer Research Periphery, Budapest, Hungary
e-mail: fleischer.kutatasi.periferia@gmail.com

with higher population density, and thus contribute to the just transition mechanism of the European Green Deal (European Commission, 2021). Additionally, it can serve as an efficient and environmentally friendly instrument in a region's Sustainable Urban Mobility Plan (SUMP) or an emerging Sustainable Regional Mobility Plan (SRMP) promoted by the European Commission's Sustainable and Smart Mobility Strategy and under the Horizon Europe programme and the CIVITAS Initiative. Our recommendations also aim to contribute to wide-scale activities in the regional cooperation programmes, such as Interreg, and other current and future initiatives, e.g., the European Rural Mobility Network (ERMN) or the European Network for Rural Development (ENRD) as part of the European CAP Network.

Technology has become increasingly relevant for DRT over time. First, the booking and dispatching system was manual, where passengers typically contacted a central dispatch office via phone to request a ride. As of today, many DRT systems use computer-assisted resource allocation and optimization, digitalized booking platforms, and initial applications of artificial intelligence. However, the transport engineering problem of how relatively low and varying demand in rural contexts can be served by passenger transport services is only one side of the coin. Most of these areas have been sparsely populated and affected by ageing, business inactivity, unemployment, and population decline. Thus, mobility services are crucial to avoid social exclusion by providing equitable mobility and access to all. It is important to note that relying solely on private cars for mobility due to the lack of quality public transport is environmentally unsustainable. How DRT can be a part of a set of solutions to maintain or even increase the population, economic activity, and sustainability in rural areas is a question of tailored regional, social, and transport policies.

Therefore, in this chapter, we address rural mobility and DRT as a potential solution from the viewpoints of both engineering (transport) and social sciences (economic and policy aspects) to provide holistic considerations for future policies. We integrate the viewpoint of traffic engineers, focusing on operational issues and efficiency, with the approach of social scientists, who are more concerned about the broader societal impact of transport systems and their effect on social inclusivity. Together, we have been able to look at DRT in a holistic manner, initially determining the policy objectives this technology should promote in a transport system given current technological and technical possibilities.

In addition to reviewing existing literature, we participated at workshops, interviewed experts, and conducted field study visits[1] to better understand current European DRT system practices. In this chapter, we first examine how DRT schemes have evolved, introducing the different business models within which these systems operate. Then, we focus on the policy goals these systems aim to achieve. Finally, we offer policy recommendations for creating DRT systems that better fulfil the policy objectives.

DRT Systems and Their Business Models

Traditional forms of public transport function effectively in areas with larger population density and a significant intensity of economic activity, making them suitable to serve environments with high built density as well. However, when examined from a financial standpoint, the construction and operation of such systems are typically not profitable, which is a legitimate reason to provide them as public services. In other words, an economic surplus can only be shown when the external effects of car use as a substitute and the additional community obligations of public transport are considered. Due to these marginal conditions, high-quality public transport is only available to a specific segment of the population.

While in theory it is a noble idea that public services of equal quality should be available to everyone regardless of their location, this condition often cannot be met in practice. Instead, as a thought experiment, one may consider another criterion of justice: everyone should receive roughly the same support in terms of public funding to address their mobility needs. This may imply that individuals residing in more challenging locations receive proportionally less support. Consequently, a larger personal contribution is inherent in their choice of residence. Even though the outcome of this thought experiment is easily debatable from a normative perspective, we observe that it is very much in line with the actual policy choice of many societies in Europe.

The question arises as to what technical solution can be employed to ensure that a public transport system is suitable for the circumstances and beneficial for the community within the given framework of public support. This is where demand responsive transport (DRT) comes to the forefront, as it can be more efficient in providing the public transport service in areas where demand is lower than what is ideal for traditional public transport.

The number of DRT services has increased for several reasons in recent decades. One reason is the already mentioned technological progress, which has introduced new forms of communication systems. With the widespread adoption of smartphones and mobile internet connectivity, booking and payment have become seamless. Through these technologies, operators can easily communicate with drivers, and route updates can be sent to them after being recalculated in response to a new rider's request. Another reason is users' attraction to on-demand services in other aspects of their lives, coupled with the increasing acceptance of the sharing economy concept. A third reason is that mobility service providers aim to enhance the efficiency of their operations, while competent authorities in Europe strive to provide public transport services to people living in sparsely populated areas. Although it is asserted that DRT is the optimal solution in these areas, they often exist in urban areas as well, providing a complementary service to traditional public transport in time or space, e.g., during night hours. These rural and urban DRT services in Europe are typically financially compensated (in line with the EC Regulation 1370/2007).

Several DRT schemes were initially established based on grants or financial support, however, many of them were later deemed too expensive by the operator, leading to the cessation of service. One such programme was the Rural Bus Challenge in the United Kingdom, which supported DRT schemes throughout the country. Although this programme played a significant role in developing DRT services in areas where conventional public transport had previously been withdrawn (Enoch et al., 2004), many of these systems ceased to exist when the subsidy expired.

One key measure that can help bring costs down is to keep the system as simple as possible. A door-to-door service with a fully flexible route is very costly to operate, and it might make it harder for potential passengers to understand how the service works in case originally they had experience only with fixed schedule—fixed route services, because the more variable travel time can mean extra burden in route planning (Currie & Fournier, 2020; Enoch et al., 2004). However, in case travel time and precise arrival time is not of high consideration, the door-to-door solution can be a simpler one, since the passenger basically does not have to do any route planning. As an alternative, stops can be predefined and travellers can request a ride to any stop, and the bus will take them to that stop on the shortest route given the demand present at other stops that might require some detour. DRT systems can be also semi-flexible, having a predefined route and running only at times and sections where demand emerges. A major trade-off related to the efficiency of these services is between the average waiting time of travellers and the size of the fleet or, in other words, where DRT should be on the scale between traditional public transport and a fully flexible taxi service.

As part of our fieldwork and interviews with experts we came to the conclusion that a DRT system is more likely to be successful in rural areas if it is operated as a many-to-many or a many-to-one service, meaning they are more likely to have high enough usage rates to get enough support from decision-makers to keep financing its losses over an extended period of time. Such services have no fixed route, nor a predefined schedule, rather the operation of the buses is organised based on the current demand, so the routing of the bus changes from day to day (Nelson & Wright, 2021). Another key factor for success is that DRT networks may operate in connection with each other in a larger area, so spare bus capacity can be reallocated and used in a neighbouring area.

What Policy Goals Do These Systems Try to Achieve?

The primary goal of DRT services is to improve the quality of life of individuals by increasing access to jobs and services, reducing car dependency in rural areas and increasing the quality of public transport. Increasing access to jobs and services can stop the decline of the rural population, as the cost of commuting (both financial and time) would decrease, and individuals would be better off by staying at or moving to locations with lower real estate prices. However, the spatial sprawl of activities and residential areas should be avoided, due to its negative external effects through the extensive use of energy and the reduction of green space.

In the absence of reliable public transport connections, rural residents must rely on individual car use. Authorities can aim to provide a certain level of public transport service for every citizen if their destination is not too remote, making it possible to replace some car trips with DRT ones, and therefore reducing carbon emissions and the financial burden of travel on rural residents. Sihvola et al. (2012) examined the needs of car users and analysed how a DRT system could improve their living conditions, although in an urban and not in a rural environment. They found that the two main justifications of personal car use were insufficient public transport connections and the need for temporal flexibility. DRT could respond to both needs.

The quality of public transport can be improved in multiple dimensions: frequency can be higher than in the case of traditional services, reducing users' schedule delay costs (the cost of having to wait for a service or having to rearrange one's activities according to the schedule); walking distances to and from the stops can be shorter, and operational hours can be extended.

One further policy goal of some special DRT services is to increase the mobility of vulnerable people, such as older people or persons with disabilities, who do not have access to personal cars and therefore are limited in their travel options. Indeed, it has been shown that different public service schemes generate unequal welfare gains and losses for different income groups (Hasnine & Habib, 2020). These schemes try to reduce transport poverty and disadvantage, empowering these people to reach better jobs and services without the need to invest in the ownership of a private car. As the least wealthy may not be able to finance the upfront cost of moving to larger cities and they cannot afford car ownership and use, DRT can be a measure against social exclusion (Knierim & Schlüter, 2021; Lucas, 2004; Lucía & Ingrida, 2022; Shergold & Parkhurst, 2012).

At some locations, tourism-related mobility can be served most efficiently via DRT services (Matsuhita et al., 2022; Ngamsirijit, 2015). This can be true either because the attractions or the accommodation are dispersed geographically, or because there are rather large variations in demand over time (i.e., due to weather).

Besides these goals, DRT can help decrease the financial cost of public transport provision via the improvement of efficiency.

Note that many of these policy goals listed above can be in conflict with each other. As an example, a DRT system with high service frequency that could be attractive to car users would have to be heavily subsidised and would not provide much benefit for the environment compared to private car use (Schasché et al., 2022). It is also important to take those into account who would be negatively affected by a DRT system, i.e., because of a larger variance in travel time.

There is already extensive research on how routes can be optimised and what this new, fully automated booking and travel assignment system is going to mean in terms of different aspects of transportation (i.e.

Amirgholy & Gonzales, 2016). The appearance of self-driving vehicles is going to make this system even more flexible and efficient, as some constraints can be excluded from the route generation algorithm, e.g., drivers' compulsory rest periods (Militão & Tirachini, 2021; Winter et al., 2018; Zhang et al., 2021). However, there might be downsides; vehicle miles travelled without carrying a passenger could, for example, increase, resulting in higher volumes of traffic (Oh et al., 2020; Winter et al., 2016).

Policy Recommendations

Local and regional transport is regulated at the national and regional levels in the EU, hence the European Commission cannot directly influence how DRT systems operate and integrate into transport networks. However, we believe that through various policy instruments (i.e., SUMP guidelines, funding programme requirements) the Commission could promote the creation of DRT systems that fit well into the region's or nation's larger transport strategy and system, are cost-effective and sustainable in the long run. The following recommendations address policies that fall into the competencies of the national or regional authorities, but the Commission could initiate them to help the long-term development of DRT in rural areas.

Our first policy recommendation is motivated by the obvious challenge of identifying the right societal goals that DRT operations can support. Previous pilot projects suggest that DRT is not a viable option under a threshold level of spatial demand density, that is, in overly sparse rural regions where occupancy rates rarely necessitate more than the capacity of a regular taxi service. The scale economies that regular scheduled bus services feature with larger vehicles and straight routes impose an upper bound on the demand intensity in which DRT is an optimal tool. However, these lower and upper thresholds are context specific values determined by the local operating costs, user preferences, and the policymakers' definition of social fairness. We recommend that policymakers should acknowledge that DRT is not a generic solution to transport provision and define explicitly the demand intensity that they intend to serve with this mode of transport.

Second, let us recall that the fundamental idea behind DRT is that it offers a flexible solution to serve low-demand markets with public transport. Assuming that this goal can be achieved, it has the potential to reduce the cost of residing in rural areas substantially. In other words, DRT has the potential to make the rural lifestyle more attractive, thus providing an incentive for a wider segment of society to follow this lifestyle, inducing rural sprawl. This may lead to a vicious circle in which an effective response to a known challenge in transport policy might further expand the scale of this challenge. Similarly, DRT should not become a competitor of scheduled public transport services as long as the latter offers a more efficient alternative through scale economies, instead, traditional public transportation and DRT should be integrated, i.e., the latter operating as a feeder service to the former. In summary, our recommendation is that effective DRT policies should be implemented in combination with complementary incentive mechanisms that maintain the benefits of concentration (i.e., agglomeration), both in terms of land use and transport flows. This supposes that DRT policies are well integrated into effective transport and regional policies.

Third, a critical part of a DRT system is financial sustainability, especially in the case of rural DRT systems, since they rely more heavily on public funding. After a period of time in which there is extended financial support to cover upfront costs, DRT systems are often considered by the operators too expensive and shut down soon after the initial source of funding exhausts. To avoid this outcome, the share of public funding should be kept at a reasonable level right in the planning phase of the project, proportional to the goals that DRT ought to achieve. Research shows that reducing the cost of operations is fundamental, and novel technologies in e-booking, ticketing, and fleet management can help achieve this. Keeping the system simple offers another opportunity: a door-to-door service with a fully flexible route is very costly to operate, and it might make it harder for potential passengers to plan a trip and understand how the system works (Currie & Fournier, 2020; Enoch et al., 2004).

Finally, we would like to highlight the fact that the entry to the DRT market is remarkably simple from a technological point of view: such services can be provided by regular minibuses that can be owned and operated with a moderate investment and a basic driving licence. This opens up the possibility of a more flexible, mixed market structure as compared to the often fully publicly owned model of regular bus services. One may envisage a model in which a transport agency operated an online platform for DRT but the actual vehicles might be provided by

SMEs, local institutions (e.g., schools) and others who own minibuses that reach a predetermined standard. This market structure, similar to the well-known business model of ride-sourcing providers, would allow for more flexibility, a better utilisation of collective transport capacities in rural regions, and reduce the fixed investment need of the authority responsible for DRT provision. The regulatory and taxation framework has to support this flexible participation in a DRT system for service providers.

In sum, the main takeaway from our chapter is that decision-makers should first define clear policy goals that they would like DRT to support. In this framework of objectives, they should define the lower and upper threshold of demand intensity that they intend to serve with DRT, and they should commit to provide the funding necessary to keep the service operational in the long run. Financial sustainability could also be supported by lowering the barriers to entering the DRT systems for service providers. DRT policies should also include incentives that support economic activities and the concentration of land use.

Note

1. We attended workshops in Milan (European Transport Conference, 2023), Kraków (Central European Excellence in Transportation Research Association, CEETRA), Brussels (European Conference of Transport Research Institutes), and Delft (Delft University of Technology). Additionally, we organised workshops in Budapest (KTI Institute for Transport Sciences and Logistics) and conducted study tours and expert interviews in Hungary, Belgium, the Netherlands, and Scotland.

References

Amirgholy, M., & Gonzales, E. J. (2016). Demand responsive transit systems with time-dependent demand: User equilibrium, system optimum, and management strategy. *Transportation Research Part b: Methodological, 92*, 234–252. https://doi.org/10.1016/j.trb.2015.11.006

Brake, J., Nelson, J. D., & Wright, S. (2004). Demand responsive transport: Towards the emergence of a new market segment. *Journal of Transport Geography, 12*(4), 323–337. https://doi.org/10.1016/j.jtrangeo.2004.08.011

Currie, G., & Fournier, N. (2020). Why most DRT/Micro-Transits fail—What the survivors tell us about progress. *Research in Transportation Economics, 83*, 100895. https://doi.org/10.1016/j.retrec.2020.100895

European Commission. (2021). *The European Green Deal*. https://commission.europa.eu/strategy-and-policy/priorities-2019-2024/european-green-deal_en

Enoch, M., Potter, S., Parkhurst, G., & Smith, M. (2004). INTERMODE: Innovations in demand responsive transport: final report. *Social Research in Transport (SORT) Clearinghouse, 22.*

Hasnine, M. S., & Habib, K. N. (2020). Transportation demand management (TDM) and social justice: A case study of differential impacts of TDM strategies on various income groups. *Transport Policy, 94*, 1–10. https://doi.org/10.1016/j.tranpol.2020.05.002

Knierim, L., & Schlüter, J. C. (2021). The attitude of potentially less mobile people towards demand responsive transport in a rural area in central Germany. *Journal of Transport Geography, 96*, 103202. https://doi.org/10.1016/j.jtrangeo.2021.103202

Lucas, K. (2004). *Running on empty: Transport, social exclusion and environmental justice*. Policy Press. https://doi.org/10.51952/9781847426000

Lucía, M. D., & Ingrida, M.-B. (2022). Transport Poverty: A systematic literature review in Europe. *Publications Office of the European Union*. https://doi.org/10.2760/793538

Matsuhita, S., Yumita, S., & Nagaosa, T. (2022). A proposal and performance evaluation of utilization methods for tourism of a demand-responsive transport system at a rural town. *2022 IEEE 25th International Conference on Intelligent Transportation Systems (ITSC)*, 2920–2925. https://doi.org/10.1109/ITSC55140.2022.9922311

Militão, A. M., & Tirachini, A. (2021). Optimal fleet size for a shared demand-responsive transport system with human-driven vs automated vehicles: A total cost minimization approach. *Transportation Research Part a: Policy and Practice, 151*, 52–80. https://doi.org/10.1016/j.tra.2021.07.004

Nelson, J. D., & Wright, S. (2021). Flexible transport services. In *The routledge handbook of public transport* (pp. 224–235). Routledge.

Ngamsirijit, W. (2015). Demand responsive transportation for creative tourism logistics planning. *International Journal of Intelligent Enterprise, 3*(1), 38–53. https://doi.org/10.1504/IJIE.2015.073446

Oh, S., Seshadri, R., Azevedo, C. L., Kumar, N., Basak, K., & Ben-Akiva, M. (2020). Assessing the impacts of automated mobility-on-demand through agent-based simulation: A study of Singapore. *Transportation Research Part a: Policy and Practice, 138*, 367–388. https://doi.org/10.1016/j.tra.2020.06.004

Regulation (EC) No 1370/2007 of the European Parliament and of the Council of 23 October 2007 on Public Passenger Transport Services by Rail and by Road and Repealing Council Regulations (EEC) Nos 1191/69 and 1107/70, 315 OJ L (2007). http://data.europa.eu/eli/reg/2007/1370/oj/eng

Schasché, S. E., Sposato, R. G., & Hampl, N. (2022). The dilemma of demand-responsive transport services in rural areas: Conflicting expectations and weak user acceptance. *Transport Policy*, *126*, 43–54. https://doi.org/10.1016/j.tranpol.2022.06.015

Shergold, I., & Parkhurst, G. (2012). Transport-related social exclusion amongst older people in rural Southwest England and Wales. *Journal of Rural Studies*, *28*(4), 412–421. https://doi.org/10.1016/j.jrurstud.2012.01.010

Sihvola, T., Jokinen, J.-P., & Sulonen, R. (2012). User needs for urban car travel: Can demand-responsive transport break dependence on the car? *Transportation Research Record*, *2277*(1), 75–81. https://doi.org/10.3141/2277-09

Winter, K., Cats, O., Correia, G. H. de A., & van Arem, B. (2016). Designing an automated demand-responsive transport system: Fleet size and performance analysis for a campus–train station service. *Transportation Research Record*, *2542*(1), 75–83. https://doi.org/10.3141/2542-09

Winter, K., Cats, O., Correia, G., & van Arem, B. (2018). Performance analysis and fleet requirements of automated demand responsive transport systems as an urban public transport service. *International Journal of Transportation Science and Technology*, *7*(2), 151–167. https://doi.org/10.1016/j.ijtst.2018.04.004

Zhang, L., Chen, T., Yu, B., & Wang, C. (2021). Suburban demand responsive transit service with rental vehicles. *IEEE Transactions on Intelligent Transportation Systems*, *22*(4), 2391–2403. https://doi.org/10.1109/TITS.2020.3027676

Open Access This chapter is licensed under the terms of the Creative Commons Attribution 4.0 International License (http://creativecommons.org/licenses/by/4.0/), which permits use, sharing, adaptation, distribution and reproduction in any medium or format, as long as you give appropriate credit to the original author(s) and the source, provide a link to the Creative Commons license and indicate if changes were made.

The images or other third party material in this chapter are included in the chapter's Creative Commons license, unless indicated otherwise in a credit line to the material. If material is not included in the chapter's Creative Commons license and your intended use is not permitted by statutory regulation or exceeds the permitted use, you will need to obtain permission directly from the copyright holder.

CHAPTER 7

Providing State-Supported Financial Incentives and Benefits for Vehicle Insurance Policies Using Telematics

Virginia Petraki, Apostolos Ziakopoulos, Evangelia Fragkiadaki, Nikolaos Karouzakis, Konstantinos Kakavoulis, and George Yannis

Abstract We recommend providing state-supported financial incentives and benefits for vehicle insurance policies using telematics. To achieve this policy recommendation, we propose the following: (1) Provision for financial incentives and benefits by the state for vehicle insurance policies using telematics across the European Union member states; (2) Conduct comprehensive social cost–benefit analysis (CBA) to assess policy

---

V. Petraki · A. Ziakopoulos · G. Yannis (✉)
Department of Transportation Planning and Engineering, National Technical University of Athens, Athens, Greece
e-mail: geyannis@central.ntua.gr

V. Petraki
e-mail: vpetraki@mail.ntua.gr

A. Ziakopoulos
e-mail: apziak@central.ntua.gr

© The Author(s) 2024
I. Keseru et al. (eds.), *Strengthening European Mobility Policy*,
https://doi.org/10.1007/978-3-031-67936-0_7

feasibility, either at a European Union or at a national level; (3) Advocate for European Union-level policy implementation supported by a centralized fund to promote telematics via insurance policies, aligning with the EU Green Deal and Vision Zero targets; and (4) Showcase benefits of interdisciplinary collaboration involving experts from transportation engineering, economics, psychology, and law for policy design and evaluation.

Keywords Telematics · Vehicle insurance · Social cost–benefit analysis

INTRODUCTION

Climate change, environmental degradation, energy use, and road safety are key existential threats to Europe and worldwide that should be addressed. Transport is responsible for about a quarter of the EU's total CO_2 emissions, 71.7% of which come from road transport (European Parliament, 2023). As an additional detrimental transport externality, road safety emerges as a major public health issue that requires immediate coordinated efforts and effective prevention, as crashes are the leading cause of death until 29 years globally. Although several efforts are being made to improve road safety, at a global level the death toll remains very high, estimated at 1.19 million fatalities annually (WHO, 2023) and at 20,640 fatalities in the EU in 2022 (EC, 2023). Therefore, the need for a solution that can mitigate these challenges is evident .

E. Fragkiadaki
School of Social Sciences, University of the West of England, Bristol, UK
e-mail: eva.fragkiadaki@uwe.ac.uk

N. Karouzakis
Alba Graduate Business School, The American College of Greece, Athens, Greece
e-mail: nkarouzakis@alba.acg.edu

K. Kakavoulis
Digital Law Experts, Athens, Greece
e-mail: k.kakavoulis@dle.gr

Undeniably, the public investments on infrastructure and the interventions in the legislation that have been carried out in Europe in the last years, have significantly contributed to the improvement of road safety; yet further contribution will be limited, and it requires high investment and significant time. Additionally, considering (i) the average vehicle age in EU (~12 years) and the old fleet in several EU countries, especially in southeast Europe with an average age greater than 15 years (ACEA, 2023), (ii) the slow renewal rate, and (iii) the relatively small role of the vehicle in crashes (Singh, 2018); the future contribution of the vehicle improvement in road safety and climate change is expected to be low. On the other hand, *driving behaviour* is the most critical factor and the root of the problem in road safety (Singh, 2018), energy efficiency, and the environment (Singh & Kathuria, 2021). Therefore, state initiatives and policies should focus on the improvement of driving behaviour, to achieve the target of decreasing road fatalities by 50% until 2030, with the most effective tool for driving behaviour assessment and improvement being telematics technology.

Telematics utilizes Artificial Intelligence and data from smartphones, devices installed in the vehicle (e.g., OBD: On-Board Diagnostics, cameras) and connected vehicles to monitor, evaluate and improve driving behaviour, promoting safe and eco-driving, reducing road crashes by 20%-50% (Reimers & Shiller, 2019; Ziakopoulos et al., 2022) and fuel consumption and CO_2 emissions by up to 30% (Barkenbus, 2010; Wu et al., 2011; Tulusan et al., 2012; Toledo & Shiftan, 2016; Michelaraki et al., 2020).

Meanwhile, insurance companies have already *integrated telematics into their insurance products*, offering Usage-Based Insurance schemes such as Pay-How-You-Drive, offering financial rewards to drivers based on their safe driving behaviour. The widespread adoption of telematics through insurance products holds the potential for significant benefits to society by reducing road crashes and the environmental impact, to consumers, and insurance companies. For consumers, telematics-based insurance offers significant advantages over traditional insurance, including education features to improve their driving behaviour, financial benefits, and rewards (e.g., discounted insurance premiums, gamification rewards, loyalty schemes). Insurance companies benefit from the ability to accurately quantify driving risk, reduce their claims costs by providing financial incentives to safe drivers and improving driving behaviour, increase their customer portfolio by providing discounted insurance premiums, create new revenue streams, increase the customer retention, and consequently decrease their Loss Ratio.

However, the vehicle insurance sector has very low capabilities of further investments and risk (Combined Ratio ~ 100%), due to low insurance premiums and relatively high crash frequency (EMIM, 2019). Therefore, state support is required for the wide acceptance and use of telematics that will lead to significant societal and environmental benefits.

In this framework, *the provision of financial incentives and benefits by the State for vehicle insurance policies using telematics* is proposed across the EU member states. The policy recommendation consists of the following:

- *Financial incentives in the form of a "Safe Pass" Voucher*: Provision of a "Safe Pass" Voucher for drivers upon the purchase of a telematics insurance policy. Alternatively, the financial incentive could be provided in the form of the complete abolition of premium tax on vehicle insurance policies using telematics in combination with a Voucher.
- *Additional Benefits for Safe Drivers*: Provision of additional benefits to safe drivers that have a score higher than a high threshold. Safe drivers who renew their insurance policy using telematics will enjoy additional benefits, including: (a) free access in city centres, (b) free parking (in areas that there is a parking cost), and (c) use of bus lanes.

This policy recommendation calls upon national governments and EU policymakers to implement initiatives such as the "Safe Pass" Voucher, incentivizing the adoption of safer and eco-friendlier driving behaviours. Insurance industry stakeholders are poised to play a key role by integrating telematics into their products, leveraging the recommended financial incentives to enhance policy attractiveness. Simultaneously, telematics industry stakeholders will be essential in providing the technological backbone, ensuring that such policies are grounded in reliable data and advanced analytics to foster safer and more environmentally friendly driving practices across Europe.

To emphasise the socio-economic feasibility of this policy, a comprehensive *social CBA case study in Greece* is conducted. Implementing such a policy requires concerted collaborative efforts across various fields, including transportation engineering, computer science, finance, behavioural psychology, digital law, and legislation related with the protection of personal data. These efforts will ensure the comprehensive design of the policy, its successful implementation, and socio-economic feasibility.

Transportation engineers and financial experts play a pivotal role in the development of the socio-economic analysis framework with advanced tools for data handling, modelling, and simulation, quantifying and projecting the long-term impact of telematics on road safety, travel time, fuel consumption, and emissions, translating these effects into monetary terms, and incorporating the appropriate economic indicators (e.g., Net Present Value (NPV)) to ensure the social feasibility of the policy. Their collaboration is integral in investigating suitable forms of financial incentives.

Behavioural psychology experts along with transportation engineers contribute significantly to comprehending (a) how individuals perceive and respond to policy changes by aiding in the development of stated preference surveys that investigate public acceptability of the proposed policy and (b) how the feedback, financial incentives, and benefits contribute to the driving behaviour improvement.

Legal experts guide the endeavour through the regulatory landscape, addressing the legal aspects of telematics technology and ensuring compliance with existing legislation and regulations related to insurance, data privacy, and consumer protection. With regard to the management of the personal data of drivers who will be insured using telematics, EU through the European Data Protection Board has already issued Guidelines regarding the management of data from Connected Vehicles, "Guidelines 1/2020 on processing personal data in the context of connected vehicles and mobility-related applications" (Adopted: March 2021) which contain specific references and examples for telematics insurance products. Therefore, the treatment of personal data in insurance policies using telematics is sufficiently regulated and it does not include any risks.

The Case Study

To demonstrate the socio-economic feasibility of the policy recommendation, a social CBA is conducted in Greece, as a case study, with a time horizon up to the year 2030, focusing on passenger cars (Petraki et al.,

2024). The analysis aims to showcase the tangible socio-economic benefits, resulting from reduced road casualties, fuel consumption, and CO_2 emissions. The methodology is based on European guidelines for CBA (EIB, 2013; Sartori et al., 2014).

Four alternative Scenarios are investigated, with different provided financial incentives and benefits by the Greek State to the insurance policies using telematics. These Scenarios S1, S2, S3, and S4 involve Safe Passes with values of €50, €55, €60, €70, considering the average car insurance premium in Greece (Insurancemarket, 2022). Scenario Zero (S0) represents the baseline "do-nothing" situation in which the provision of Safe Pass is not considered and against which, Scenarios S1-S4 are compared. The aforesaid values of Safe Passes are considered indicative for Greece to achieve high demand of insurance telematics, and they should be adjusted in any other country, considering the average vehicle insurance premium.

A *questionnaire survey* was conducted (Petraki et al., 2024), using the stated preference methodology, to determine the level of public acceptance of each Scenario. Behavioural psychology expertise contributed in the survey by ensuring the questionnaire format was unbiased and concise, and adhered to ethical standards. From the 1,250 respondents, the answers of 897 passenger car drivers were considered. Based on the answers a linear regression model is developed to predict the sensitivity of the public acceptability of telematics against the financial incentives showing that the percentage of drivers who would buy a telematics insurance policy depends significantly on the financial benefits provided (p-value ≤ 0), highlighting the need for the state to provide financial incentives and benefits to telematics insurance policies.

Therefore, considering the number of insured vehicles in Greece (HAIC, 2023) and the policy acceptance, the annual number of Safe Passes for each Scenario in Table 7.1 is expected that will be fully consumed. Anticipating a period for updating and maturing the market and the implementation period, the number of Safe Passes offered in the first year (2024), is lower by 50–60% compared to the subsequent years.

For each Scenario, the estimated State Grant for the provision of the Safe Pass, and *the effects* on road casualties, fuel consumption, travel time, and CO_2 emissions from passenger vehicles by 2030, are calculated and expressed in monetary units. Specifically:

Road Safety

- Injury crash statistics in Greece of 2019 (prior to COVID-19 pandemic) are considered as representative, including road fatalities, severe and light injured road users in the category of passenger car.
- The social costs per road fatality, severe and light injury are valued at 2,148,034€, 273,574€, and 51,373€, respectively, in Greece (ITF, 2020).
- An average 30% reduction in road casualties is assumed, based on literature (as cited in the Introduction).

Fuel Consumption

- The average annual fuel consumption (litres per vehicle-kilometre) for the Greek passenger car fleet by 2030 is considered, based on EU targets (Yang & Bandivadekar, 2017).
- Fuel consumption effect is estimated, considering the fuel cost, the annual vehicle-kilometres travelled on Greek roads, and the average fuel consumption.
- An average 5% reduction in fuel consumption is assumed, based on literature (as cited in the Introduction).

Travel Time

- The travel time effect is estimated considering the number of insured passenger cars in Greece, an average car occupancy rate of 1.2 (Eurostat, 2023), the annual travel time on Greek roads, and a value of travel time (VOT) at 5.6€/hour (EC, 2019; Eurostat, 2021).
- The potential increase in travel time because of the speed reduction resulting from the improvement in driving behaviour was conservatively considered equal to 2% (Kontaxi et al., 2021).

Environment

- The environmental effect was computed considering the annual vehicle-kilometres travelled, the CO_2 emissions per vehicle-kilometre, and the social cost of CO_2 (€/tonne) (EIB, 2020; EC, 2021).

- An average 5% reduction in CO_2 emissions is assumed, based on the international literature (as cited in the Introduction).

State Grant

- The State Grant is estimated for each Scenario, considering the Safe Pass value and the number of Safe Passes in each Scenario.

Considering the required State Grant and the socio-economic effects of the implementation of the recommended policy, the Internal Rate of Return (IRR), the present value of economic benefits (PV), and the NPV are estimated. The costs and benefits arising at different times are discounted using the Social Discount Rate (SDR) which is considered equal to 0.8% (EC, 2021) (Table 7.1).

Table 7.1 Social CBA for the implementation of telematics insurance policies in Greece

Scenarios	S1	S2	S3	S4
Safe Pass Value	50€	55€	60€	70€
Annual Safe Pass Offer	0.7 mil	1.5 mil	2.5 mil	3.5 mil
Total State Grant (2024–2030)	−225 mil€	−533.5 mil€	−960 mil€	−1,575 mil€
State Grant (2024)	15 mil€	38.5 mil€	60 mil€	105 mil€
Annual State Grant (2025–2030)	35 mil€	82.5 mil€	150 mil€	245 mil€
Change in socio-economic indicators (2024–2030)				
Light Injured (persons)	−1,331	−2,841	−4,669	−6,560
Severe Injured (persons)	−62	−131	−219	−307
Fatalities (persons)	−75	−158	−261	−364
Fuel Consumption (litres)	−121 mil	−270 mil	−450 mil	−636 mil
CO_2 Emissions (tonnes)	−0.3 mil	−0.6 mil	−1 mil	−1.5 mil
PV (0.8%)	320 mil€	685 mil€	1,134 mil€	1,590 mil€
NPV (0.8%)	100 mil€	164 mil€	197 mil€	55 mil€
IRR	52.7%	35.3%	24.3%	4.8%

Note 2024 indicators multiplied by 75% due to the policy application post the first quarter

It is concluded that in all Scenarios, there is a significant reduction in road injuries and fatalities and a significant environmental benefit. Specifically, the positive NPV and the high IRR (5% < IRR < 53%), *indicate the socio-economic feasibility of the policy recommendation in Greece for all examined Scenarios*. In case that the main criterion is the minimization of the State Grant, Scenario S1 is the preferred one. In terms of socio-economic performance, S3 is the preferred one as it demonstrates the highest NPV and a high IRR index (24.3%). In case that the main criterion is the maximisation of the social and environmental impact, which is the main motivation for this policy, S4 is the preferred one. Extrapolating the results of the CBA in Greece, to EU and all vehicles, the recommended policy could result to 740–4,440 less road fatalities per year in EU, depending on the level of the financial incentives.

It is highlighted that the recommended policy refers to all vehicles, whereas the CBA refers only to passenger vehicles; therefore, the potential societal and environmental benefit could be even higher. Also, *the current methodology can be applied to other countries* intending to adopt this policy recommendation by adjusting key parameters such as Safe Pass values, average vehicle insurance premiums, public acceptability through stated preference surveys, social cost per road casualty, VOT, and other relevant factors.

CONCLUSION AND RECOMMENDATIONS

This chapter underscores the *critical importance of addressing road safety, climate change, and, energy consumption* as pressing global challenges. This can be achieved for the transport sector via the promotion and wide use of vehicle telematics through the provision of financial incentives and benefits by the State for vehicle insurance policies using telematics.

The provision of financial incentives and benefits by the State for vehicle insurance policies using telematics is proposed across the EU member states. The recommended policy introduces a new innovative approach in road safety, that is mainly based on safe behaviour, and not in the traditional, until now, approach of punishment.

To *assess the socio-economic feasibility of this policy*, a comprehensive social CBA was conducted, with a focus on a case study in Greece. Four alternative scenarios, each offering different levels of financial incentives, were examined, along with a "do-nothing" scenario as a baseline reference point. The results highlight that in all Scenarios there are significant societal and environmental benefits, with a significant reduction in road casualties, fuel consumption, and CO_2 emissions, as well as positive socio-economic indicators. It is important to acknowledge that, while CBA provides valuable insights, its results are subject to underlying uncertainties and assumptions, especially given the long-term analysis time horizon. Therefore, conducting sensitivity analysis is crucial for robust policy formulation and decision-making. Naturally, updating the CBA forecasts as the time-period runs and adding more detailed layers as they become available is a fruitful way to navigate any uncertainty.

Collaboration across diverse disciplines is essential for the design, implementation, and evaluation of the proposed policy recommendation. An integrated approach, involving experts in transportation engineering, economics, psychology, and legal disciplines, is pivotal for defining financial incentives, developing a social CBA to assess the policy feasibility, measuring societal response, and ensuring compliance with EU privacy regulations.

More thorough and tailored upscaling of the present findings across an EU level can potentially be more fruitful than present results at a national level. Specifically, *the provision of financial incentives and benefits for vehicle insurance policies using telematics should be adopted as a policy at the EU level*, to maximise the societal and environmental benefits. Such policy should be accompanied by CBA studies, either at an EU or at a national level, considering the societal, environmental, and macroeconomic indicators of the EU members in order to define the optimum value of financial incentives (Safe Pass) and benefits per country.

This EU policy should also be supported by a centralised EU fund that will be available for each country (on the top of any national funds that may be provided by each country) and it will be specifically dedicated to the promotion of telematics via insurance policies so that EU can achieve its targets related with road crashes by 2030. In summary, the provision of financial incentives and benefits in the insurance policies using telematics serves as a strategic approach for all EU members, aligning with the EU Green Deal and seamlessly contributing to Vision Zero targets, and promoting sustainable mobility.

References

Barkenbus, J. N. (2010). Eco-driving: An overlooked climate change initiative. *Energy Policy, 38*(2), 762–769.

European Automobile Manufacturers' Association (ACEA) (2023). *Vehicles in use report*. acea.auto/files/ACEA-report-vehicles-in-use-europe-2023.pdfacea. auto/files/ACEA-report-vehicles-in-use-europe-2023.pdf

European Commission (EC) (2019). *4.K83—Handbook on the external costs of transport—January 2019*. https://publications.europa.eu/resource/cellar/e021854b-a451-11e9-9d01-01aa75ed71a1.0001.01/DOC_1

European Commission (EC) (2021). *Economic appraisal vademecum 2021–2027*. www.ec.europa.eu/regional_policy/en/information/publications/guides/2021/economic-appraisal-vademecum-2021-2027-general-principles-and-sector-applications

European Commission (EC) (2023, October 19). *Road safety: 20,640 people died in a road crash last year—progress remains too slow*. transport.ec.europa.eu/news-events/news/road-safety-20640-people-died-road-crash-last-year-progress-remains too slow-2023-10-19_en

European Investment Bank (EIB) (2013). *The economic appraisal of investment projects at the EIB*.

European Investment Bank (EIB) (2020). *The EIB group climate bank roadmap 2021–2025*. eib.org/en/publications/the-eib-group-climate-bank-roadmap

European Motor Insurance Markets (EMIM) (2019). *European motor insurance markets*. https://insuranceeurope.eu/publications/465/european-motor-insurance-markets/

European Parliament (2023, February 14). CO_2 emissions from cars: facts and figures. https://europarl.europa.eu/news/en/headlines/society/20190313STO31218/co2-emissions-from-cars-facts-and-figures-infographics

Eurostat (2021, February 11). *Distribution of distance travelled per person per day by travel purpose for urban mobility on all days Table 3 Feb 2021*. ec.europa.eu/eurostat/statistics-explained/index.php?title=File:Distribution_of_distance_travelled_per_person_per_day_by_travel_purpose_for_urban_mobility_on_all_days_Table_3_Feb_2021.png

Eurostat (2023, September 11). *Passenger mobility statistics*. ec.europa.eu/eurostat/statistics-explained/index.php?title=Passenger_mobility_statistics

Hellenic Association of Insurance Companies (HAIC) (2023). *Statistical yearbook for motor insurance 2022*. http://www1.eaee.gr/sites/default/files/statisticalyearbook2022.pdf

Insurancemarket (2022). *Report 2022*. https://d23vfvjlkuzpf0.cloudfront.net/wp-content/uploads/sites/3/2023/03/IM-infographic-report_2022.pdf

International Transport Forum (ITF) (2020). *Road Safety Report, Greece, OECD*. itf-oecd.org/sites/default/files/greece-road-safety.pdf

Kontaxi, A., Ziakopoulos, A., & Yannis, G. (2021). Trip characteristics impact on the frequency of harsh events recorded via smartphone sensors. *IATSS Research, 45*(4), 574–583.

Michelaraki, E., Kontaxi, A., Papantoniou, P., & Yannis, G. (2020). *Correlation of driver behaviour and fuel consumption using data from smartphones.* Proceedings of the 8th Transport Research Arena TRA 2020 Conference (Helsinki, Finland, 27–30 April 2020).

Petraki, V., Ziakopoulos, A., Fragkiadaki, E., Yannis, G., Kakavoulis, K., & Karouzakis, N. (2024). *Social cost-benefit analysis.* Zenodo.

Reimers, I., & Shiller, B. R. (2019). The impacts of telematics on competition and consumer behavior in insurance. *The Journal of Law and Economics, 62*(4), 613–632.

Sartori, D., Catalano, G., Genco, M., Pancotti, C., Sirtori, E., Vignetti, S., & Del Bo, C. (2014). *Guide to cost-benefit analysis of investment projects.* Economic appraisal tool for Cohesion Policy, 2020.

Singh, H., & Kathuria, A. (2021). Profiling drivers to assess safe and eco-driving behavior–A systematic review of naturalistic driving studies. *Accident Analysis & Prevention, 161,* 106349.

Singh, S. (2018). *Critical reasons for crashes investigated in the national motor vehicle crash causation survey.* (Traffic Safety Facts Crash Stats. Report No. DOT HS 812 506). Washington, DC: National Highway Traffic Safety Administration.

Toledo, G., & Shiftan, Y. (2016). Can feedback from in-vehicle data recorders improve driver behavior and reduce fuel consumption? *Transportation Research Part a: Policy and Practice, 94,* 194–204.

Tulusan, J., Staake, T., & Fleisch, E. (2012, September). Providing eco-driving feedback to corporate car drivers: what impact does a smartphone application have on their fuel efficiency? In *Proceedings of the 2012 ACM conference on ubiquitous computing* (pp. 212–215).

World Health Organization (WHO) (2023, December 13). *Road traffic injuries.* https://www.who.int/news-room/fact-sheets/detail/road-traffic-injuries

Wu, C., Zhao, G., & Ou, B. (2011). A fuel economy optimization system with applications in vehicles with human drivers and autonomous vehicles. *Transportation Research Part d: Transport and Environment, 16*(7), 515–524.

Yang, Z., & Bandivadekar, A. (2017). *Light-duty vehicle greenhouse gas and fuel economy standards.* The international Council on clean Transportation.

Ziakopoulos, A., Petraki, V., Kontaxi, A., & Yannis, G. (2022). The transformation of the insurance industry and road safety by driver safety behaviour telematics. *Case Studies on Transport Policy, 10*(4), 2271–2279.

Open Access This chapter is licensed under the terms of the Creative Commons Attribution 4.0 International License (http://creativecommons.org/licenses/by/4.0/), which permits use, sharing, adaptation, distribution and reproduction in any medium or format, as long as you give appropriate credit to the original author(s) and the source, provide a link to the Creative Commons license and indicate if changes were made.

The images or other third party material in this chapter are included in the chapter's Creative Commons license, unless indicated otherwise in a credit line to the material. If material is not included in the chapter's Creative Commons license and your intended use is not permitted by statutory regulation or exceeds the permitted use, you will need to obtain permission directly from the copyright holder.

PART IV

New Methods and Tools to Support Participatory Planning

CHAPTER 8

Assessing Mobility Policy with AI-Driven Analysis of User-Generated Content

Floriano Tori, *Charlotte van Vessem*, *Juliana Betancur Arenas, and Vincent Ginis*

Abstract We recommend assessing mobility policies with AI-driven analysis of user-generated content. To achieve this policy recommendation, we propose to take into account the following: (1) Using large language models to analyse user-generated content is a reliable methodology for gathering and analysing typically overlooked relevant information regarding citizens' perceptions in the implementation of sustainable mobility policies; (2) The substantial processing capacity of these models, coupled with their ability to gather a great amount of information, enables

---

F. Tori (✉) · C. van Vessem · J. Betancur Arenas · V. Ginis
Vrije Universiteit Brussel, Brussels, Belgium
e-mail: floriano.tori@vub.be

C. van Vessem
e-mail: Charlotte.van.vessem@vub.be

J. Betancur Arenas
e-mail: juliana.betancur.arenas@vub.be

V. Ginis
e-mail: Vincent.ginis@vub.be

© The Author(s) 2024
I. Keseru et al. (eds.), *Strengthening European Mobility Policy*,
https://doi.org/10.1007/978-3-031-67936-0_8

decision-makers to supplement and enhance the often-limited traditional data collection methods. As exposed in the following case study, this methodology can provide historical perceptual information on transport modes, mobility policies, and infrastructure, among others; and (3) The ease of applying this methodology through AI open-source recent developments such as ChatGPT allows decision-makers and their teams to rapidly generate and assess a great amount of relevant data. This can facilitate policymakers' effectiveness and efficiency in the decision-making processes in urban mobility planning. However, policymakers should be aware of the characteristics of their selected population and use this as a complementary and evaluative method.

Keywords Large language models · Twitter · Mobility planning · Policy interventions

Introduction

The continuous growth demand for the European transport sector, which emits major externalities such as greenhouse gas emissions, air- and noise pollution, congestion, land uptake, and traffic accidents (UNECE, 2015), is proving to be one of the major bottlenecks in the EU's ambition to become the first climate neutral continent. In this regard, the Green Deal and the Fit for 55 proposals package serve as a roadmap for constructing and updating the EU policy framework for shifting towards a sustainable, safe, and clean mobility system. This chapter addresses policy recommendations for the EU Green Deal—specifically, those dealing with the Sustainable Smart and Mobility Strategy to accelerate this digital and green shift. Despite being a milestone in public policy, this action plan fails to include concrete actions towards already sustainable and accessible modes like active and shared mobility systems (Kwasniok & Bolmer, 2021). Moreover, when implementing part of these measures in the urban context through Sustainable Urban Mobility Plans (SUMPs), in some cases, citizens' and industries' responses have taken the form of strong backlash (Pettersson et al., 2021).

Such strong citizens' reactions can be partially due to decision-makers' lack of understanding of the population's needs. One way to identify the demands, opinions, and experiences of the people affected by these transformations is the use of participatory processes (Pourhashem et al., 2021). However, there are still major challenges to achieving large-scale participation during these processes. Some of the weaknesses of this method are the lack of data, the unawareness of appropriate communication tools, and the need for time-consuming and resource-intensive activities (Nared, 2020). Furthermore, transport planning relies heavily on traditional data collection methods such as manual surveys and traffic counts, which are expensive, tedious, and prone to errors (Zannat & Choudhury, 2019). These quantitative data collections fail to consider multidimensional aspects of travel behaviour, leading to the wrong assumption that the travel experiences of one group are universal and resulting in blind spots regarding characteristics such as gender, ethnicity, age, or income. In order to overcome these obstacles and "unlock" the modal shift potential, there is a need for innovative and interdisciplinary methods, to understand users' mobility perceptions and willingness to change. In this regard, digitalisation plays an increasingly prominent role in our mobility system (European Environment Agency, 2023). While traffic system optimisation and modelling are traditional ways to improve mobility systems, emerging technologies like big data and artificial intelligence allow for the use of new methods in innovative ways, which can enable positive behavioural shifts. Digital technologies can provide greater data availability and sophisticated assessing systems, complementing traditional research methods by extending the data scope (European Environment Agency, 2023).

Additionally, Large Language Models (LLMs), such as GPT-4, that process data in the form of natural language (i.e., produced by humans as opposed to binary language produced by computers) provide an opportunity to gain an understanding of citizens' perceptions and feelings regarding changes in active mobility on a significantly larger scale than manual analysis. Using these LLMs, it is possible to perform wide-scale sentiment analysis on user-generated content (UGC), such as social media. Sentiment analysis is a task whose goal is to analyse individuals' opinions, attitudes, and emotions towards concepts such as products, services, organisations, locations, and events (Liu, 2015), which can encompass many approaches. For our purpose, we focus on classifying text polarity (i.e., positive, neutral, or negative) towards a specific transport mode generated by Twitter users. Using sentiment analysis on UGC has the advantage of analysing populations that are otherwise difficult to reach with traditional data collection methods. Nonetheless, this recent methodology and its scope have been under-researched in mobility.

This research provides a novel complementary, evaluative, and easy-to-use methodology for sustainable transport planning that allows: (i) the collection of qualitative posts on a large scale, creating a quantitative database of these posts, (ii) that can then be analysed using GPT models to be sorted based on sentiment. This approach combines quantitative STEM methods for the data collection with social science behavioural analysis, including a gender-sensitive regard, to test its use in a case study and analyse the possible advantages it could bring to EU decision-makers. Therefore, in this chapter, we explore the usability of natural language models for analysing UGC (such as tweets) to understand the public's perceptions regarding different transport modes. This interdisciplinary and collaborative approach between STEM (data analytics) and SSH (mobility behaviour studies) aims to provide a replicable and accessible methodology to collect qualitative data using some recent developments in AI.

Evidence Case

1. Data Collection

Using Twitter (now known as X) as a source can be motivated due to multiple aspects. Due to its low cost, ease of access, and presence of public figures (Naseem et al., 2021), Twitter offers an approachable way to voice concerns or praise about policies. Additionally, tweets also contain information about location and time, which offer additional dimensions for analysis.

We collected tweets discussing mobility in the Brussels Capital Region, Belgium, between 2017–2022. A tweet was collected if it contained a previously defined mobility keyword (for example: metro) and either the name of a (local) politician, a neighbourhood or municipality, or a (shared) mobility provider. Combining keywords with these conditions allowed the widest net possible to be cast while limiting false positives (i.e. tweets being collected which are not relevant). Since the target of the collection is tweets produced by users, we excluded accounts from media institutions, automated accounts, and political parties. Due to the multilingual element of Brussels, we considered tweets in Dutch and French (the region's two official languages) and English (accounting for the large population of non-Belgian natives). Searches were done on what is called "Parent Tweets", i.e. original tweets of a user. For each parent tweet, we also collected the conversation of the tweet, which consists of "Reply Tweets", namely tweets that reply to the original tweet or to a reply tweet of the original tweet, or "Quote Tweets", which are tweets that repost the parent tweet on the user's timeline together with a comment.

After collecting tweets, we filtered out those recognised as irrelevant, delivering a total of 15,001 tweets, shown in Fig. 8.2, panel I (the dataset is publicly available online (Tori et al., 2024)). A first remark is the large presence of reply tweets, which comprise 74% of the dataset. We also note that peaks in the distribution can be correlated with known mobility interventions. For instance, the peaks observed between July and December 2022 (Fig. 8.2, panel I) correspond with the implementation of the low traffic measures regulated on the Brussels SUMP, which caused great citizen resistance and media exposure (POLITICO, 2022).

2. Assessing GPT-4 as an Expert Annotator

The second part of our research was to assess the potential of GPT-4 to serve as an expert annotator using a random sample of 500 tweets from the dataset. For this subset, three experts (two experts in urban mobility from SSH and one expert in Machine Learning, specifically in the context of natural language processing from a STEM background) manually annotated the tweets. As the aim is to understand public perception and feelings regarding the sustainable transition to active and shared modes of transport, tweets were labelled in three categories: (i) The mode of transport being mentioned, (ii) The domain in which the mode is mentioned (iii) The sentiment regarding this mode (see Fig. 8.1 for details).

This sample was then also labelled in the same categories by GPT-4. We used GPT-4 by giving it the collected tweets together with specific instructions about the task it needs to perform. This type of labelling in different categories simultaneously is made possible because GPT-4 is a generative model, meaning it can generate any text in response to an instruction.

An evaluation of the GPT-4 labelling was made by comparing the agreement on the labels between expert annotators and GPT-4. This

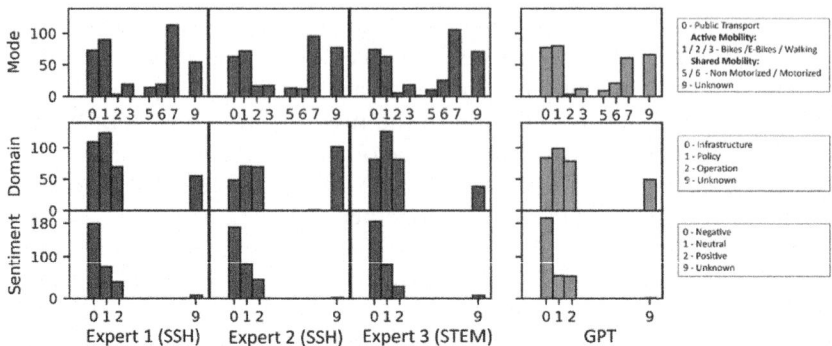

Fig. 8.1 Distribution of the labelled tweets by expert annotators (blue) and by GPT-4 (red) (color figure online)

agreement is quantified by computing the Cohen kappa (Cohen, 1960), a score between 0 and 1. The higher the value, the higher the level of agreement between annotators. A value between 0.41–0.6 denotes a moderate agreement, and between 0.61–0.8 a strong agreement. The average Cohen Kappa between the expert annotators for the sentiment of the tweets was 0.54, indicating a moderate agreement. When only considering the parents-type tweets, this agreement climbed to 0.61. This indicates that labelling reply tweets were more difficult, probably due to the absence of context (reply tweets are collected in a random ordering). When computing the annotator agreement between GPT-4 and the three experts, we find again an average coefficient of 0.54 on all tweets and 0.61 on parent tweets. This indicates an equivalent agreement among the experts as between experts and GPT-4. This implies that GPT-4 could be used instead of experts to label and understand sentiments expressed in mobility-related tweets. Figure 8.1 displays the labelling distribution of all three categories of the three experts and GPT-4, showing a similar distribution for all four as expected.

3. Using GPT-4 to Assess Collected Tweets

Building upon our previous analysis, which assessed GPT-4 as an equivalent expert annotator, we used GPT-4 as a case study on the larger dataset of tweets collected for this study. Of the 15,001 tweets collected, 8182 were considered relevant by GPT-4. Tweets were considered relevant if the "mode" and the "domain" category were not both labelled with a "9" (i.e. unknown). Tweets being attributed both a 9 for "mode" and "domain" occurred most often for reply tweets of (very) long conversations, which indicates the difficulties of understanding the context of a long conversation. Afterwards, tweets were considered further if they did not contain the label "9" in any category (i.e. the tweet mentioned a mode, a domain for this mode, and a sentiment regarding this). This narrowed the pool of tweets further down to 5059. The distribution of the labels of each category for these 5059 selected tweets can be seen in Fig. 8.2, panel II. From this, we see that the three modes most discussed were *active mobility: bikes*, *motorised mobility*, and *public transport*.

From the labelling produced by GPT-4 we can observe a few elements. First is the notable variation in the number of tweets across different transportation modes. For instance, while non-motorised shared mobility only appeared in a total of 25 tweets, bikes & e-bikes received together over 2000 tweets. In each group, we looked at the distribution of the tweets containing a certain domain, from which we observed that the majority of tweets are associated with infrastructure discussions; on the sentiment side, we note that a negative sentiment was attributed to tweets more often. Important to note, however, is that tweets labelled with a negative sentiment do not all contain opposition to existing infrastructure; these tweets also express a negative sentiment due to the desire for better infrastructure.

Conclusion and Recommendations

Emerging AI developments allow for access to larger amounts of data, which can be used to analyse mobility behaviour. Sentiment analysis can become a complementary source of information to traditional data collection sources of mobility perceptions, such as travel surveys. Moreover, LLMs proved to be as reliable as human experts when performing classification and sentiment labelling. Additionally, the automated labelling by LLMs speeds up the data collection process and can be done in multiple languages simultaneously.

UGC is an accessible and relatively unexplored source of mobility information for policymakers and planners. As shown in Fig. 8.2, panel I, UGC peaks correlate with major mobility policy interventions. This information allows decision-makers to trace the impact of their policies in a specific

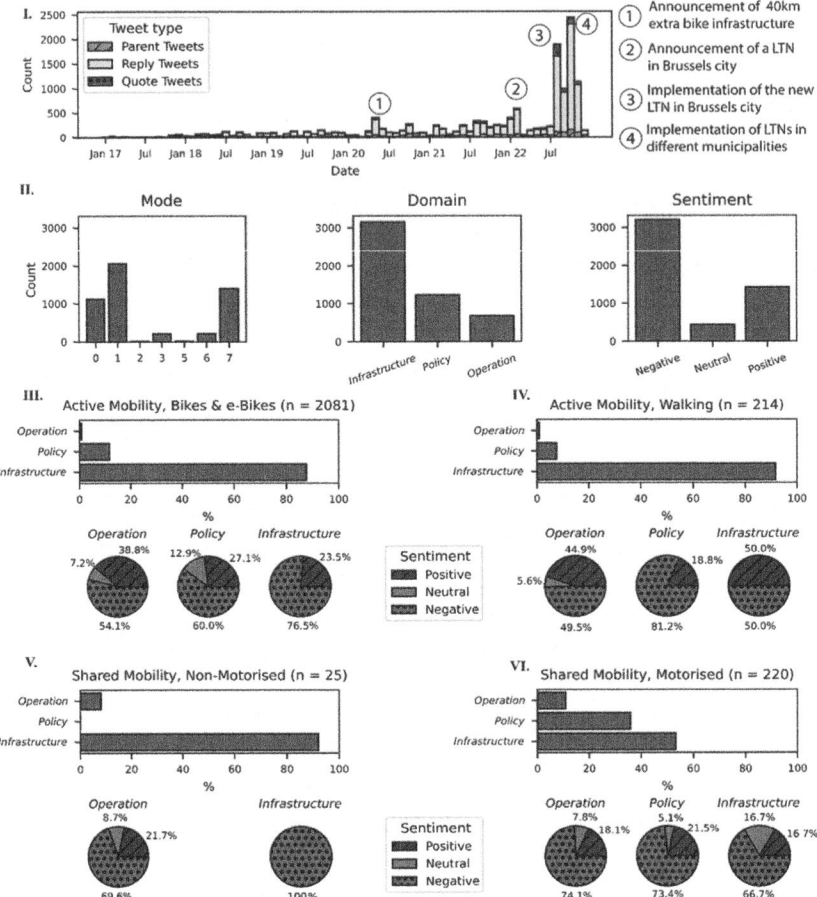

Fig. 8.2 **I.** Distribution by date of creation of the total tweets collected. Peaks in the distribution can be correlated to known mobility interventions. The label indicates the difference in tweet types. **II.** Aggregated histograms of tweets labelled by GPT-4 for each labelling category (Mode mentioned, domain, and sentiment). **(III, IV, V, VI).** For each active and shared mobility mode we selected the tweets labelled by GPT-4 to contain that mode. For each of these subcategories of tweets, we show the distribution of the domain mentioned (in percentages of total tweets of the selected mode), and within the domains, we show sentiment percentages with pie charts

timeframe. Additionally, our methodology's capacity to analyse sentiments towards transport modes and their specific domain can provide evaluative feedback for new interventions and issues to be addressed. However, while UGC can provide data from a wider group of people, it should not be viewed as a way to access the opinion of an entire population, just as sentiment analysis cannot provide a population-wide consensus on an intervention. Rather, this method is a complementary evaluative method to traditional outreach and analysis. Special attention should be given to the information source. As previously shown, the number of replies exceeds the number of parent tweets, constituting 74% of the dataset. Therefore, we recommend considering these replies when using UGC as a data source.

The usage of publicly available AI tools makes this method accessible to people without a data science background. This opens possibilities to study sentiments quickly and in detail, for example, by zooming in on a specific mode or area. We attempted to test this ourselves by analysing the link between mobility-related UGC and gendered language in our dataset but found very few results. This might have been due to the sampling order of primarily sampling tweets concerning mobility and performing a second search on gender in these posts. For a more thorough analysis, we suggest reversing this process or pairing the keywords on mobility and gender into one step of the sampling process.

The main limitations of our methodology are based on the inter-annotator agreement process, especially regarding reply tweets, which is something GPT-4 also struggles with when labelling long conversations. Our methodology also highlights that language sentiment might not be equal to the mobility sentiment expressed (i.e. sarcasm or irony). Secondly, despite the broad scope that this methodology can have, it is important to highlight that its results only represent part of the population, excluding those who do not have access to these platforms or any other digital media. As demonstrated, the methodology exhibits a wide range of applications and potential for conducting diverse analyses, which could be replicated in other cities. In the case of Brussels, infrastructure discussions concerning active modes such as bikes and e-bikes have been central to recent debates, while walking has received comparatively less attention. Shared mobility, both motorised and non-motorised, is even less represented, possibly due to its recent introduction and adoption in cities.

Considering the rapid advancements in AI, the use of this methodology is promising for the future and has great improvement potential and future research should focus on the application of this model on other UGC platforms.

References

Cohen, J. (1960). A coefficient of agreement for nominal scales. *Educational and Psychological Measurement, 20*(1), 37–46. https://doi.org/10.1177/001316 446002000104

European Environment Agency. (2023). *Transport and environment report 2022 Digitalisation in the mobility system: Challenges and opportunities* (EEA Report No 07/2022). https://www.eea.europa.eu/publications/transport-and-environment-report-2022/transport-and-environment-report/view

Kwasniok, R., & Bolmer, A.M. (2021). *The sustainable and smart mobility strategy of the European commission—A critical assessment.* https://www.changing-transport.org/wp-content/uploads/EU-Mobility-Strategy.pdf

Liu, B. (2015). *Sentiment analysis: Mining opinions, sentiments, and emotions* (1st ed.). Cambridge University Press. https://doi.org/10.1017/CBO978 1139084789

Nared, J. (2020). Participatory transport planning: The experience of eight European metropolitan regions. In J. Nared & D. Bole (Eds.), *Participatory research and planning in practice.* Springer International Publishing. https://doi.org/10.1007/978-3-030-28014-7_2

Naseem, U., Razzak, I., Khushi, M., Eklund, P. W., & Kim, J. (2021). COVIDSenti: A large-scale benchmark twitter data set for COVID-19 sentiment analysis. *IEEE Transactions on Computational Social Systems, 8*(4), 1003–1015. https://doi.org/10.1109/TCSS.2021.3051189

Pettersson, F., Stjernborg, V., & Curtis, C. (2021). Critical challenges in implementing sustainable transport policy in Stockholm and Gothenburg. *Cities, 113*, 103153. https://doi.org/10.1016/j.cities.2021.103153

POLITICO. (2022, November 22). *Brussels wrestles with local anger over plans to curb traffic.* POLITICO. https://www.politico.eu/article/brussels-local-anger-good-move-plan-curb-traffic-elke-van-den-brandt/

Pourhashem, G., Malichová, E., Kovacikova, T., & Sk. (2021). *The role of participation behavior and information in nudging citizens sustainable mobility behavior: A case study of Bratislava region.* https://doi.org/10.1109/ICETA5 4173.2021.9726681

Tori, F., Betancur, A. J., Ginis, V., & van Vessem, C. (2024). Brussel mobility Twitter sentiment analysis CSV dataset. *Zenodo.* https://doi.org/10.5281/zenodo.11401124

UNECE. (2015). *Transport for sustainable development—The case of inland transport.* https://unece.org/transport/publications/transport-sustainable-development-case-inland-transport

Zannat, K. E., & Choudhury, C. F. (2019). Emerging Big Data Sources for Public Transport Planning: A Systematic Review on Current State of Art and Future Research Directions. *Journal of the Indian Institute of Science, 99*(4), 601–619. https://doi.org/10.1007/s41745-019-00125-9

Open Access This chapter is licensed under the terms of the Creative Commons Attribution 4.0 International License (http://creativecommons.org/licenses/by/4.0/), which permits use, sharing, adaptation, distribution and reproduction in any medium or format, as long as you give appropriate credit to the original author(s) and the source, provide a link to the Creative Commons license and indicate if changes were made.

The images or other third party material in this chapter are included in the chapter's Creative Commons license, unless indicated otherwise in a credit line to the material. If material is not included in the chapter's Creative Commons license and your intended use is not permitted by statutory regulation or exceeds the permitted use, you will need to obtain permission directly from the copyright holder.

CHAPTER 9

Enabling Inclusive Urban Transport Planning Through Civic Artificial Intelligence

Dimitris Michailidis, Kristina Khutsishvili, Konstantinos Konstantis, Aristotle Tympas, Imad Antoine Ibrahim, and Sennay Ghebreab

Abstract We recommend enabling inclusive urban transport planning through civic artificial intelligence. To achieve this policy recommendation, we propose the following: (1) Encourage and provide resources for experimentation with new technologies that enable local community participation in urban transport planning; (2) Recognize the potential of Artificial Intelligence (AI) to assist in complex urban transport planning decisions; (3) Acknowledge that AI is embedded in society, instead of

D. Michailidis (✉) · S. Ghebreab
Socially Intelligent Artificial Systems, Informatics Institute, University of Amsterdam, Amsterdam, The Netherlands
e-mail: d.michailidis@uva.nl

S. Ghebreab
e-mail: s.ghebreab@uva.nl

K. Khutsishvili
Informatics Institute, University of Amsterdam, Amsterdam, The Netherlands

treating it as a neutral technology; and (4) Foster community engagement in transport planning and evaluation, via a Civic AI framework that directly integrates preferences and feedback into planning.

Keywords Artificial intelligence · Public transport · Participatory planning

INTRODUCTION

Transport networks, such as bus and metro lines, are the foundation of urban living (Martens, 2016). Investing in expanding public transport lies at the core of the European Union's Green Deal initiative, which aims at reducing transport emissions towards a climate neutral Europe. However, planning new or expanding existing transport networks requires overcoming physical, socio-economic, political, and legal challenges. This represents a problem that demands innovative solutions. Recent advancements in Artificial Intelligence (AI) have opened up possibilities for understanding and addressing some of these challenges (Michailidis et al., 2023), making urban transport an area where AI, combined with policies, can be pivotal in contributing to the European Green Deal (Fetting, 2020).

Artificial Intelligence was initially perceived as aiming at constructing computers with equal (or superior) mental capacities to the human

---

K. Konstantis · A. Tympas
Department of History and Philosophy of Science, National and Kapodistrian University of Athens, Athens, Greece
e-mail: konstkon@phs.uoa.gr

A. Tympas
e-mail: tympas@phs.uoa.gr

I. A. Ibrahim
Faculty of Behavioural, Management and Social Sciences, University of Twente, Enschede, The Netherlands
e-mail: i.ibrahim@utwente.nl

brain (Simos et al., 2022). By now, however, AI is connected to the so-called "smartness mandate", a social drive to make everything, from individual artefacts to broader networks and infrastructures, smart (Halpern et al., 2017).

Critical histories of technology suggest that AI, like all other technologies, is not neutral. Social, economic, and political interests are advanced through certain AI configurations (Garvey, 2018; Simos et al., 2022). In response, the field of AI Ethics has emerged, alongside special studies from the interdisciplinary field of Science and Technology Studies (STS), which aim at opening the black box of AI. STS focuses on biases that may be advanced through the opaque design of AI (Burrell, 2016; Pasquale, 2015). Comparatively, AI Ethics focuses more on the issues emerging during the use of AI after the design stage—with privacy, fairness, and accountability being some of them (Müller, 2020). Alongside STS and AI Ethics, additional concerns are being studied, including the hidden labour required to operate AI (O'Neil, 2016; Pasquinelli, 2023).

Within this context, AI has been specifically presented as a solution to transport-related challenges (Dia, 2023). Here too, issues regarding the opaqueness and unethical use of AI have arisen, as they may perpetuate social biases. Other issues are related to privacy and safety (European Commission: Directorate-General for Research and Innovation, 2020), and issues of accountability and responsibility (Blackett, 2022). Here, we instead focus on the problem of planning public transport lines, a problem in which considerable benefits can emerge from ethical and participatory usage of AI.

Many of the challenges faced by AI systems today stem from their globally oriented, top-down governance. A prime example involves language models, like ChatGPT, which are controlled by big corporations with the necessary resources for data collection and training (Schneier & Waldo, 2023). This process poses significant risks in terms of power concentration and ethical issues of these systems (Konstantis et al., 2023).

In this chapter, our interdisciplinary team, comprising researchers from AI, STS, economics, law, and human rights backgrounds, presents a framework for a locally oriented, bottom-up approach to transport planning policy (Forum for the Future, 2017). Our aim is to propose a method that involves communities in the decision-making process. Initially, we examined recent AI models proposed for transport planning. While these models offer significant potential to aid planners in making better decisions, we observed a predominant top-down approach in their development, neglecting the input of affected communities. Subsequently, we analysed large commercial models like ChatGPT and Gemini, identifying two techniques applicable to local transport planning: Reward Shaping and Reinforcement Learning from Human Feedback. Adapting these techniques requires ensuring representative and fair participation. To gain insights, we investigated participatory methodologies used in other fields, particularly in STS and economics. Through iterative discussions and knowledge exchange, we identified essential aspects to expand upon in our policy recommendation. We integrated these insights with our expertise to develop a framework for Civic AI-based transport planning, emphasising community engagement and inclusivity. We cover the technical components of the system, methodologies for fostering community engagement in a safe, collaborative environment, and potential European-level legislation to support its implementation. In the subsequent section, we elaborate on these key aspects.

With this chapter, we aim to encourage the European Commission to support research and experimentation with innovative technologies—such as Civic AI—that enable local community inclusion in decision-making processes. Through technology that integrates preferences and feedback into the process, we emphasise the potential for AI to contribute to the European Green Deal initiative, by promoting more sustainable and inclusive urban transport networks.

Transport Network Planning

Transport network planning involves an authority that decides where to build new transport networks or expand existing ones. Traditionally, it follows a process where, firstly, the future travel demand of a city is forecasted based on predicted demographics and economic activity. Subsequently, the current network is evaluated, identifying areas that may face capacity challenges. Following this assessment, potential projects of new lines are proposed and evaluated to determine their ability to meet the forecasted demand, alleviate congestion, and fit within the available budget. Finally, a shortlist of qualified projects is extracted and planned (Martens, 2016).

This process does not address a fundamental dimension: the fair distribution of benefits of the new lines. Furthermore, it follows a strict top-down approach, ignoring the input of affected urban communities. By emphasising efficiency, it aims to alleviate congestion, neglecting other fundamental considerations, such as environmental sustainability and access to opportunities (hereby referred to as accessibility) (Martens, 2016). New planning concepts have emerged to address the gap, considering evaluations based on factors such as CO_2 emissions and the number of accessible facilities.

Introducing additional factors into decision-making creates two challenges. First, planning transport projects becomes increasingly complex. Second, given the needs of different communities, determining the relative importance of the different factors for prioritisation is difficult. Hence, one possible solution is to develop AI-driven systems to facilitate informed decisions considering the needs of different communities.

Civic Artificial Intelligence

Our policy framework is grounded in Civic Artificial Intelligence (Civic AI), which promotes the participation of citizens in public decisions using AI (Duberry, 2022). Civic AI applies the Civic Studies framework to the challenges of rapid technological development and revolves around the fundamental question: "What should we do?" (Levine, 2022; Ostrom, 1990). In Civic AI systems, citizens are not passive consumers of technology, or mere data points to be used in models; rather, they are included as co-designers, actively contributing to their creation (Hsu et al., 2022). To achieve this, AI systems should be advanced from a globally oriented, top-down, to a locally oriented, bottom-up process.

Local community involvement can be achieved in different ways, including participatory design workshops that shape research questions, community-led data collection, or survey-based system evaluation (Hsu et al., 2022). By actively engaging with AI systems, citizens enhance their technological proficiency, while researchers gain a better understanding of societal requirements. Ultimately, this paradigm cultivates greater confidence among citizens that their needs are being addressed (Hsu et al., 2022).

In the prevailing top-down planning paradigm, users are treated as data-generation artefacts. For instance, recent AI-based tools employed for predicting future mobility demand rely on mobile phone GPS data to estimate current movements (Michailidis et al., 2023). These data sources are notorious for exhibiting biases against those who do not own latest-technology phones (Coston et al., 2020). In contrast, in our proposed bottom-up approach, citizens actively engage in the planning and evaluation of the system's output. Specifically, the AI agent learns to adapt its behaviour to user preferences. Users actively assess the agent's output throughout the training process, ensuring their involvement at every stage.

Bringing together community members and technological providers is difficult and requires actions in bridging the gap in terminology and facilitating co-creation. Translational activities can prepare communities for professional interactions with technological providers for a mutually rewarding engagement, with insights into social innovation tools and challenges (Khutsishvili, 2024). Ethical considerations come to the forefront, especially when engaging vulnerable and marginalised groups (such as older people, disabled, refugees, low-literate people), emphasising the "do no harm" principle (Khutsishvili et al., 2024). During co-creation, the biggest risk is community disappointment, stemming from factors like unaffordable final solutions, inability to immediately benefit from the solution, or feeling undervalued or unheard. Expectation management is a crucial tool to mitigate these risks. Initially, it is important to acknowledge the asymmetry of information and power disparity, prioritising translational activities and motivational frameworks aimed at balancing the co-creation setting.

Various EU rules provide the basis for the inclusion of citizens and communities in a Civic AI framework. Article 11 of the Treaty on European Union (TEU) states that institutions "shall, by appropriate means, allow citizens and representative associations to make known and publicly exchange their views in all areas of Union action" (TEU, 1992, Art. 11(1)), as well as participate in the democratic life of the union (TEU, 1992, Art. 10(3)). Additionally, one of the objectives of the Better Regulation agenda concerns involving citizens, businesses, and other stakeholders in decision-making, with the ultimate goal of enhancing the legitimacy of the democratic process (Bunea & Chrisp, 2023).

The recent framework for regulating Artificial Intelligence by the European Parliament (AI Act) forms the basis for citizen inclusion in AI systems design. Title V stipulates the creation of regulatory sandboxes for testing the new AI technology before its introduction to the market (European Commission, 2021). Article 55 provides specific measures for users and small-scale providers, including priority access to the sandbox, awareness-raising activities, and dedicated channels for communication (European Commission, 2021). Another example is the General Data Protection Regulation (GDPR) that focuses on the "protection of natural persons with regard to the processing of personal data and rules relating to the free movement of personal data" (European Parliament, Council of the European Union (2016), Art. 1). Although the GDPR does not require direct engagement of impacted parties (Skoric et al., 2022), it is a step towards fostering active involvement.

The AI Act and the GDPR represent some mechanisms adopted at the European level that include enhancing the participation of citizens, communities and civil society. In conjunction with general EU law, it shows that the union considers these stakeholders' roles, albeit further efforts are needed.

Civic AI for Transport Network Planning

In Fig. 9.1, we present the proposed framework for inclusive transport planning. Its key technical component is the transport planning agent. An agent is an AI model that iteratively learns to make decisions within a virtual environment. In the context of transport, such an environment is established by creating a grid of the city, with each area represented by a grid cell. Various layers of crucial metrics are encoded in this environment, such as forecasted travel demand, accessibility, and emissions.

The agent aims to generate transport lines that effectively balance a combination of these metrics. A fundamental AI paradigm for implementing such agents is Reinforcement Learning (RL). An RL agent learns through trial-and-error, by taking actions, receiving feedback from the

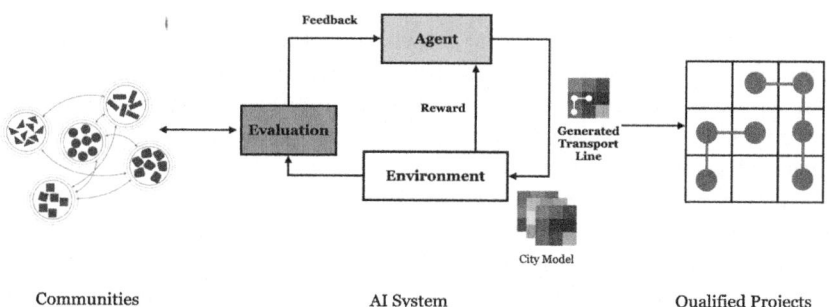

Fig. 9.1 A framework for inclusive transport planning, in which communities actively engage in the design and evaluation of the system used to generate transport projects

environment in the form of a reward, and adapting its behaviour to maximise it. Through this training process, the agent continually improves its decision-making.

For a transport planning agent, actions involve sequentially connecting areas in the city to form a transport line. Traditionally, this is automated and without interference, with domain experts deciding beforehand how to combine various metrics into a single reward. However, under the Civic AI framework, this process involves non-experts that influence the training process.

There exist various processes to incorporate community feedback into Reinforcement Learning (Kaufmann et al., 2023). We outline two classes that policymakers can utilise: Reward Shaping and Reinforcement Learning from Human Feedback (RLHF).

Reward Shaping draws from insights from social choice theory, which concerns the aggregation of individual preferences into collective decisions. In this process, communities collaborate to co-design the reward function that will train the transport planning agent. The agent thus learns to maximise the collective reward formulated by the communities.

Reward shaping occurs through direct data collection. Citizens, through an interface, provide their most important journeys or their preferred transport lines, based on their needs. The data can subsequently be utilised by experts to incorporate the relative community importance of various aspects (e.g. accessibility, emissions) into the reward function. This can lead to a reward that expresses the desirable compromise between the traditional objectives and the citizens' preferences. This is a process that enhances citizens' agency, as they are asked to directly submit their preferences. However, one drawback lies in the potential variety of the gathered data, making it challenging to reach an optimal compromise that satisfies everyone. Additionally, it requires substantial community input, making it crucial for policymakers to ensure the collection of a diverse dataset.

In Reinforcement Learning from Human Feedback (RLHF), communities are engaged throughout the entire training process, not solely during the reward-shaping phase. RLHF can be used alongside Reward Shaping, as a final tuning step. In contrast with Reward Shaping, citizens here evaluate generated lines by the agent. At various intervals during training, the agent generates alternative extensions, which are then assessed by citizens. In RLHF, this evaluation commonly takes place through direct comparison. Citizens are provided with a set of alternative lines, and they are asked to rank them from most to least preferable. This preference is then fed back to the agent to update the reward function.

This process requires fewer data and can be effectively implemented in a small, co-creative space. Additionally, it is less sensitive to specific individual needs, as the generated lines for evaluation align better with metric-based objectives. A drawback of this approach is that it limits citizens' evaluations to lines already considered good by the agent, thereby affecting their agency in the decision-making process. Nevertheless, it can lead to a meaningful compromise between the traditional planning approach and the Civic AI framework.

The training process of the agent concludes when it stops improving. This is a straightforward procedure in the system, which keeps track of the received feedback. Upon completion, the agent can be used to generate the eligible transport projects. Optionally, another layer of evaluation can be incorporated, for example via a final voting on the projects, or by expert planners.

A Hypothetical Example

Let us consider an example where a city plans to expand its public transport network by building a new metro line. In addition to traditional mobility forecasts and budget constraints, the planners decide to use the Civic AI framework to incorporate citizens' feedback. Workshops are organised to gather input from representative community members.

Citizen feedback occurs in two phases. Initially, citizens provide factors such as commuting preferences and environmental concerns. Transport planners use this input to shape the reward function of the planning agent. They aggregate preferences into factors like demand and travel ease, then use Reward Shaping to weight these factors accordingly.

The planner agent is then trained in a simulated environment to draw the metro line by maximising the given reward function. The second phase of citizen feedback begins, via Reinforcement Learning from Human Feedback, where citizens rank the generated metro lines based on their preferences. The agent adjusts its outputs to reflect these preferences. When it stops improving, qualified transport projects are extracted. These may undergo a final round of evaluation, either through participatory or expert assessment, to ensure alignment with city goals and values.

Conclusion

In this chapter, we combined our interdisciplinary expertise to propose a framework for Civic AI-based transport planning. Through discussions, knowledge sharing, and reviews, we have reached a recommendation that encompasses technical, social, and legal aspects of the framework. While practical implementation requires further experimentation, workshops are already underway across Europe to address how to incorporate diverse perspectives into decision-making. We outline the next steps required to advance the proposed framework.

Firstly, it is important to recognise Artificial Intelligence (AI) as a powerful, specialised tool rather than an omniscient, neutral technology that can take unbiased, universal decisions. By doing so, policymakers acknowledge its potential to assist in the complex decision-making process of planning public transport inclusively, while at the same time understanding its limitations.

Encouraging research and experimentation with innovative technologies that promote community inclusivity is crucial. The European Commission should allocate resources and funding towards initiatives that support the development and experimentation of AI solutions tailored to local contexts. This should focus on transitioning from a globally oriented, top-down approach to a locally oriented, bottom-up approach in decision-making. Resources could include funding pilot projects, offering technical assistance, and facilitating knowledge sharing among member states. Through these initiatives, AI can be leveraged for the benefit of local communities and to advance inclusive decision-making processes in transport.

Finally, to foster community engagement, we propose implementing processes that integrate preferences and feedback into AI systems' training and evaluation, via the methods we outlined in the chapter. By actively involving the community in the planning and evaluation of transport projects, we can ensure that the resulting infrastructure aligns with the needs and desires of those it serves, ultimately leading to more sustainable and equitable urban transport networks.

Acknowledgements The research work of Konstantinos Konstantis is supported by the Hellenic Foundation for Research and Innovation (HFRI) under the 3rd Call for HFRI PhD Fellowships (Fellowship Number: 5188).

References

Blackett, C. (2022). The ethics of AI in autonomous transport. In M. C. Leva, E. Patelli, & L. Podofillini (Eds.), Proceedings of the *32nd European Safety and Reliability Conference (ESREL 2022)* (pp. 3390–3397). Research Publishing. https://doi.org/10.3850/978-981-18-5183-4_J03-05-453-cd

Bunea, A., & Chrisp, J. (2023). Reconciling participatory and evidence-based policymaking in the EU better regulation policy: Mission (im)possible? *Journal of European Integration, 45*(5), 729750. https://doi.org/10.1080/07036337.2022.2144848

Burrell, J. (2016). How the machine 'thinks': Understanding opacity in machine learning algorithms. *Big Data and Society, 3*(1), 1–12. https://doi.org/10.1177/2053951715622512

Coston, A., Guha, N., Ouyang, D., Lu, L., Chouldechova, A., & Ho, D. E. (2020). Leveraging administrative data for bias audits: Assessing disparate coverage with mobility data for Covid-19 policy. arXiv:2011.07194 [Cs, Stat]. http://arxiv.org/abs/2011.07194

Dia, H. (Ed.). (2023). *Handbook on Artificial Intelligence and transport.* Edward Elgar.

Duberry, J. (2022). AI and civic tech: Engaging citizens in decision-making processes but not without risks. In *Artificial Intelligence and democracy* (pp. 195–224). Edward Elgar Publishing.

European Commission. (2021). *Proposal for a regulation of the European Parliament and of the Council laying down harmonised rules on artificial intelligence (Artificial Intelligence Act) and amending certain Union legislative acts* (COM/2021/206 final). https://eur-lex.europa.eu/legal-content/EN/TXT/?uri=CELEX%3A52021PC0206

European Commission: Directorate-General for Research and Innovation. (2020). *Ethics of connected and automated vehicles: Recommendations on road safety, privacy, fairness, explainability and responsibility.* Publications Office of the European Union. https://data.europa.eu/doi/10.2777/035239

European Parliament, Council of the European Union. (2016). *Regulation (EU) 2016/679 of the European Parliament and of the Council of 27 April 2016 on the protection of natural persons with regard to the processing of personal data and on the free movement of such data, and repealing Directive 95/46/EC (General Data Protection Regulation).* https://eur-lex.europa.eu/eli/reg/2016/679/oj

Fetting, C. (2020). *The European Green Deal.* ESDN Office, Vienna. https://www.esdn.eu/fileadmin/ESDN_Reports/ESDN_Report_2_2020.pdf

Forum for the Future. (2017). *Citizens bringing the future forward.* Futures Centre Issuu. https://issuu.com/futurescentre/docs/fftf_citizens_bringing_the_future_f

Garvey, C. (2018). Broken promises and empty threats: The evolution of AI in the USA, 1956–1996. *Technology's Stories, 6*(1). https://doi.org/10.15763/jou.ts.2018.03.16.02

Halpern, O., Mitchell, R., & Geoghegan, B. D. (2017). The smartness mandate: Notes toward a critique. *Grey Room, 68*, 106–129. https://doi.org/10.1162/GREY_a_00221

Hsu, Y.-C., 'Kenneth' Huang, T.-H., Verma, H., Mauri, A., Nourbakhsh, I., & Bozzon, A. (2022). Empowering local communities using Artificial Intelligence. *Patterns, 3*(3), 100449. https://doi.org/10.1016/j.patter.2022.100449

Kaufmann, T., Weng, P., Bengs, V., & Hüllermeier, E. (2023). A survey of reinforcement learning from human feedback. arXiv:2312.14925 [Cs, LG]. http://arxiv.org/abs/2312.14925

Khutsishvili, K. (2024). *Guidelines for translating frameworks, methods, tools and principles of local innovations for marginalised and vulnerable communities—2023*. Open Research Europe. https://open-research-europe.ec.europa.eu/articles/4-36

Khutsishvili, K., Pavicic, N., & Combé, M. (2024). The challenge of co-creation: How to connect technologies and communities in an ethical way. *Proceedings of the ETHICOMP 2024. 21st International Conference on the Ethical and Social Impacts of ICT*. https://dialnet.unirioja.es/servlet/articulo?codigo=9333577

Konstantis, K., Georgas, A., Faras, A., Georgas, K., & Tympas, A. (2023). Ethical considerations in working with ChatGPT on a questionnaire about the future of work with ChatGPT. *AI and Ethics*. https://doi.org/10.1007/s43681-023-00312-6

Levine, P. (2022). *What should we do? A theory of civic life*. Oxford University Press.

Martens, K. (2016). *Transport justice: Designing fair transportation systems*. Routledge.

Michailidis, D., Ghebreab, S., & Santos, F. P. (2023). Balancing fairness and efficiency in transport network design through reinforcement learning. *Proceedings of the 2023 International Conference on Autonomous Agents and Multiagent Systems* (pp. 2532–2534). https://dl.acm.org/doi/10.5555/3545946.3598992

Müller, V. C. (2020, Winter). Ethics of Artificial Intelligence and Robotics. In *The Stanford Encyclopedia of Philosophy*. https://plato.stanford.edu/archives/win2020/entries/ethics-ai

O'Neil, C. (2016). *Weapons of math destruction: How Big Data increases inequality and threatens democracy*. Crown.

Ostrom, E. (1990). *Governing the commons: The evolution of institutions for collective action*. Cambridge University Press.

Pasquale, F. (2015). *The Black Box society: The secret algorithms that control money and information*. Harvard University Press.

Pasquinelli, M. (2023). *The eye of the master: A social history of Artificial Intelligence*. Verso.

Schneier, B., & Waldo, J. (2023, May 30). Big Tech isn't prepared for A.I.'s next chapter. *Slate*.

Simos, M., Konstantis, K., Sakalis, K., & Tympas, A. (2022). "AI can be analogous to steam power" or from the "post-industrial society" to the "fourth industrial revolution": An intellectual history of Artificial Intelligence. *ICON, 27*(1), 97–116.

Skoric, V., Sileno, G., & Ghebreab, S. (2022). *Legality, legitimacy, and instrumental possibility in human and computational governance for the public sector*. https://ceur-ws.org/Vol-3289/paper4.pdf

Treaty on European Union. (1992). http://data.europa.eu/eli/treaty/teu_2012/oj

Open Access This chapter is licensed under the terms of the Creative Commons Attribution 4.0 International License (http://creativecommons.org/licenses/by/4.0/), which permits use, sharing, adaptation, distribution and reproduction in any medium or format, as long as you give appropriate credit to the original author(s) and the source, provide a link to the Creative Commons license and indicate if changes were made.

The images or other third party material in this chapter are included in the chapter's Creative Commons license, unless indicated otherwise in a credit line to the material. If material is not included in the chapter's Creative Commons license and your intended use is not permitted by statutory regulation or exceeds the permitted use, you will need to obtain permission directly from the copyright holder.

CHAPTER 10

Facilitating Sustainable Logistics Policy Development Using Multicriteria Satisfaction Analysis: A Case of Preference Mapping for Cargo Bike Last-Mile Delivery

He Huang, Xu Zhang, Salvatore Corrente, Sajid Siraj, and Maja Kiba-Janiak

Abstract We recommend facilitating sustainable logistics policy development using multicriteria satisfaction analysis. With regard to this policy recommendation, through a case study of preference mapping for cargo bike last-mile delivery we demonstrate the following: (1) The proposed

---

H. Huang (✉)
Laboratory for Energy Systems Analysis, Paul Scherrer Institute, Villigen PSI, Switzerland
e-mail: he.huang@vub.be

Mobilise Mobility and Logistics Research Group, House of Sustainable Transitions, Vrije Universiteit Brussel, Brussels, Belgium

X. Zhang
School of Transport and Civil Engineering, TU Dublin, Dublin, Ireland
e-mail: xu.zhang@tudublin.ie

© The Author(s) 2024
I. Keseru et al. (eds.), *Strengthening European Mobility Policy*,
https://doi.org/10.1007/978-3-031-67936-0_10

MUlticriteria Satisfaction Analysis (MUSA) based public perception elicitation survey tool offers an alternative approach to map public preferences in sustainable policy decision-making; (2) The findings suggests different cities have different sustainability priorities for sustainable urban freight transport; and (3) City managers and logistics practitioners could offer tailored policies and services to address citizens' needs.

Keywords Public opinions · Urban logistics · Cargo bikes

INTRODUCTION

Public Participation in Urban Logistics Policy Development

The booming of e-commerce and on-demand instant freight deliveries in the city bring changes in economic activities, consumption behaviours, demand patterns, and disruptions in mobility, which posed substantial challenges to urban logistics operations (Dablanc, 2023; Dablanc et al., 2017). The urban shipments become smaller and fragmented, resulting in an increased number of direct delivery trips to home destinations (Amling & Daugherty, 2020; Dablanc, 2019; Hopkins & McCarthy, 2016).

Urban logistics has a significant impact on the functionality of urban areas and the well-being of their citizens. In fact, 70% of the European population lives in cities and 23% of EU transport greenhouse gas emissions come from urban areas (European Commission, 2024). The European Green Deal has set its target to achieve a 90% reduction in

---

S. Corrente
Department of Economics and Business, University of Catania, Catania, Italy
e-mail: salvatore.corrente@unict.it

S. Siraj
Leeds University Business School, Leeds, UK
e-mail: s.siraj@leeds.ac.uk

M. Kiba-Janiak
Wrocław University of Economics and Business, Wrocław, Poland
e-mail: maja.kiba-janiak@ue.wroc.pl

transport-related greenhouse gas emissions by 2050, compared to 1990 levels (Tsavachidis & Le Petit, 2022). Along with the New European Urban Mobility Framework launched in 2021, the European Commission has put urban mobility and logistics in the spotlight of the policy agenda.

In urban mobility and logistics policy development, stakeholder engagement is a crucial component of Sustainable Urban Mobility Plans (SUMP) and Sustainable Urban Logistics Plans (SULP), as well as in the New EU Urban Mobility Framework. Public engagement in policymaking ensures an inclusive and effective planning process; it involves a range of activities, such as public consultation, dialogue, and participation. To develop urban logistics policies, researchers in the urban logistics field have tested various forms of public engagement with citizens, such as focus groups (Tuomala et al., 2023) and living labs (Maltese et al., 2023). Yet, public participation is still a challenging task due to limited resources and policy tools (Maltese et al., 2023), and insufficient information provided to the public to make informed choices (Tuomala et al., 2023).

Cities vary in their approaches to visioning and planning sustainable mobility, so is their key stakeholders' opinions (Foltýnová et al., 2020). The stakeholders' perception and endorsement of the sustainable mobility concept and their ability to express their views will impact on urban mobility decisions (Foltýnová et al., 2020). Therefore, there is an urgency for decision-makers to capture public's opinion and investigate citizen's preferences on last-mile delivery solutions, thus, to provide tailored policies in the local context.

To address this challenge, the Social Sciences and Humanities (SSH) research agenda emphasises the importance of interdisciplinary research to address critical societal challenges in sustainable transport and mobility development (Ryghaug et al., 2023). Researchers advocate for more inclusive and deliberative approaches in collaboration with actors and stakeholders to enhance the effectiveness of policymaking (Ryghaug et al., 2023). In response to this "SSH CENTRE" project's vision to encourage SSH-STEM collaboration, our research contributes novel perspectives from the fields of logistics and supply chain management within the social sciences and humanities (SSH) domain, as well as decision-making and mathematics within the science, technology, engineering, and mathematics (STEM) domain. The SSH researchers specialised in urban logistics in this study built a narrative scenario to facilitate the participants' understanding towards this cargo bike delivery topic. The STEM researchers specialised in group decision-making applied the Multicriteria Satisfaction Analysis (MUSA) methodology for data analysis. Meanwhile, the online questionnaire design, dissemination, contents and results were compiled together.

In this chapter, we present a novel technique for eliciting preferences to facilitate sustainable logistics policy development. We apply a multicriteria decision analysis method, i.e. MUSA, aimed at assessing the opinions of public stakeholders in urban logistics policymaking. We demonstrate its practical usefulness through an empirical study to map the mass public perceptions of cargo bike as a means of last-mile delivery. This methodology seeks to offer a comprehensive understanding of the complexities inherent in urban logistics contexts, drawing upon interdisciplinary insights from both SSH and STEM disciplines.

Research Methods and Survey Design

In this study, we propose an inclusive and intuitive *preference mapping approach* based on the MUSA method to elicit participants' preferences mapping in the policymaking process (Grigoroudis & Siskos, 2002). Mapping the preferences of the public towards services is crucial, as it enables service providers to tailor their offerings to meet the specific needs and expectations of their customers, thereby enhancing satisfaction and fostering sustained engagement (Czepkiewicz et al., 2018). In this context, the MUSA method has been used to precisely map public preferences for a specific service, illustrating its utility in capturing nuanced consumer insights. However, it is usually applied with a rather small sample size (Grigoroudis & Siskos, 2010). In this study, we implemented the MUSA method within a mass-participation scenario with over 2,000 participants.

To illustrate our approach, an evidence-based business case using a "cargo bike delivery service" was developed as a *hypothetical* scenario. Based on this scenario, we developed a MUSA-based framework to obtain and aggregate citizen feedback from multiple cities, incorporating a diverse range of socio-demographic backgrounds, for policy recommendations.

In our MUSA application, participants are asked to express their judgments, including their overall and specific satisfaction level on several criteria towards the hypothetical cargo bike delivery service. A predetermined α level ordinal satisfaction scale is used to capture these judgments, namely "extremely dissatisfied", "somewhat dissatisfied", "neither satisfied nor dissatisfied", "somewhat satisfied", and "extremely satisfied". This scale facilitates the quantification of customer feedback. In MUSA, the overall satisfaction function is represented by Y^*, while the partial satisfaction functions corresponding to each individual criterion i are denoted by X_i^*. In this study, the criteria are factors that influence citizens' satisfaction and the overall effectiveness of the cargo bike delivery service like CO_2 emissions, noise etc. The relationship between these variables is explained by an ordinal regression analysis equation:

$$Y^* = \sum_{i=1}^{n} b_i X_i^*,$$

$$\sum_{i=1}^{n} b_i = 1,$$

where b_i is the weight of criterion i. Drawing upon the specified equation, MUSA constructs a Linear Programming model designed to discern how satisfaction across multiple criteria contributes to overall satisfaction with the service with the objective of minimising estimation errors derived from participants' inputs. The MUSA output presents a comprehensive set of results, including the overall satisfaction, which is aggregated by partial satisfaction for individual criteria with the respective weights of these criteria. A series of optimisations is performed following the initial one to infer the value functions on criteria and at the global level that better represent the citizen satisfaction. This method ensures that the derived weights are robust, accurately reflecting the priorities of the participants. Additionally, MUSA yields a series of indices, which offer deeper insights, enhancing the interpretability and reliability of the satisfaction assessment results:

1. **Average Satisfaction Indices (ASI)**: Represent the mean of the global or partial value functions, normalised within the range [0,1].

The higher the value, the higher the satisfaction with the corresponding criteria. ASI is denoted as follows:

$$\text{ASI} = \frac{1}{100} \sum_{m=1}^{\alpha} p^m y^{*m},$$

$$\text{ASI}_i = \frac{1}{100} \sum_{k=1}^{a} p_i^k x_i^{*k}, \text{ for } i = 1, 2, \ldots, n,$$

where p^m and p_i^k are the frequencies of customers belonging to the overall satisfaction level and partial satisfaction levels on criterion i, respectively.

2. **Average Demanding Indices (ADI)**: The average demanding indices are normalised in the interval $[-1,1]$. If the index reaches a value of 1, it indicates that participants exhibit the highest level of demand. In this scenario, participants are only satisfied with the utmost quality level. On the other hand, an index value of -1 signifies the lowest level of demand, where participants have minimal expectations or demands from the service or product in question. ADI is denoted as follows:

$$\text{ADI} = \frac{1 - \frac{\overline{\overline{y}}^*}{50}}{1 - \frac{2}{\alpha}}, \text{ for } \alpha > 2,$$

$$\text{ADI}_i = \frac{1 - \frac{\overline{\overline{x}}^*}{50}}{1 - \frac{2}{\alpha}}, \text{ for } \alpha > 2, \text{ and } i = 1, 2, \ldots, n,$$

where $\overline{\overline{y}}^*$ and $\overline{\overline{x}}^*$ are the mean values of functions Y^* and X_i^*

3. **Average Improvement Indices (AII)**: These indices are normalised in the interval $[0,1]$. The improvement index for a given criterion is inversely proportional to its performance level, given a certain weight. Specifically, a higher weight assigned to a criterion, coupled with lower performance in that area, results in a correspondingly higher improvement index for that criterion. This relationship highlights areas requiring enhanced focus for improvement, based on their significance and current performance levels:

$$I_i = b_i(1 - \text{ASI}_i), \text{ for } i = 1, 2, \ldots, n.$$

Upon conducting a thorough literature review, we have identified the following key criteria relevant to our study. The survey questions were designed based on each criterion (as listed in Table 10.1).

Research Finding

We surveyed with a computer-assisted web interviewing method in five capital cities in Europe, namely London, Paris, Rome, Dublin, and Warsaw. These capital cities are varied in their urban "freightscape" in terms of population and employment densities (Rodrigue et al., 2017; Rose et al., 2017). Therefore, it would be interesting to explore citizens' perceptions of cargo bike delivery in different urban archetypes.

The target survey participants are the general population over 18 years old residing in these cities. All the surveys were published in English, with additional translations available in French, Italian, and Polish. The data collection was conducted from November 2023 to January 2024.

As a result, a total of 2,030 responses were obtained across five cities (Huang et al., 2024). A statistically significant difference has been observed between urban and suburban samples in London. Consequently, we applied MUSA separately to the urban and suburban samples within London. For the other cities, the MUSA was applied to the combined samples without distinction. Due to the limited size of the suburban sample in Warsaw, we proceed by analysing the entire Warsaw dataset as a single group.

Based on the geographical locations, we clustered participants' responses into 6 groups, namely urban London ($n = 528$), suburban London ($n = 185$), Paris ($n = 545$), Rome ($n = 527$), Dublin ($n = 167$), and Warsaw ($n = 78$). The clustered responses were then analysed using MUSA to derive satisfaction value functions for each area.

Citizens in all five capital cities were highly supportive of the hypothetical introduction of cargo bike delivery services, demonstrating a growing demand for environmentally friendly last-mile delivery solutions. Surveys consistently showed high levels of satisfaction, reflecting appreciation for the overall value proposition of the hypothetical sustainable cargo bike delivery service.

In the post-optimality analysis phase, we compared the performance of the cargo bike service across 5 key criteria. We calculated the associated ASIs, ADIs, and AIIs. Using these indices along with the criteria weights, we developed two types of recommendation diagrams: the *action*

Table 10.1 Key criteria and survey questions

Criterion	Code	Description and Survey Question
CO_2 emissions	c_1	Electric cargo bike delivery can reduce CO_2 emissions by 30–55% per package (Carracedo & Mostofi, 2022). Cargo bikes can significantly reduce CO_2 emissions and air pollution (such as particulate matter and nitrogen oxide) compared to fossil-fuel vehicles. By transitioning to our e-cargo bikes, we project our reduction of carbon emissions by 70–90% compared to diesel vans, and by a third compared to electric vans Q_1: How satisfied do you feel with this scenario?
Noise	c_2	Quieter operation, less noise for your parcel delivery. Our cargo bikes are designed to be especially quieter than motorcycles or mopeds, with an average of 50–60 dB of noise, making our delivery operations less disruptive in urban areas Q_2: How satisfied do you feel with this scenario?
Traffic	c_3	Cargo bikes can significantly reduce congestion in cities. Given the compact size of our cargo bikes, we anticipate a 75% reduction in the road space required for our cargo bike fleet compared to a normal car (Cairns & Sloman, 2019). Using cargo bikes will allow our package delivery workers to work more efficiently while reducing the number of motorised vehicles in the city (Llorca & Moeckel, 2021) Q_3: How satisfied do you feel with this scenario?
Safety	c_4	Improved safety for pedestrians, and less disturbance from parcel delivery activities. While ensuring the health and safety conditions for our workers using e-cargo bikes to deliver your parcel, given the slower operational speeds (maximum of 25 km/h) (Gonzalez-Calderon et al., 2022) and reduced number of delivery vans blocking roads, cycle paths and pavements, we predict a significant reduction in the number and severity of traffic accidents related to deliveries in the city Q_4: How satisfied do you feel with this scenario?
Shipping cost	c_5	Better working conditions for our riders, but slightly more expensive for your shipping cost. With the commitment to offer fair and qualitative jobs for our e-cargo bike riders, customers may expect a slight 10–20% increase in the delivery fee compared to traditional delivery services Q_5: How satisfied do you feel with this scenario?
Overall satisfaction	v	Q_6: Considering all the above-mentioned information together, how satisfied do you feel with our e-cargo bike delivery option?

and *improvement* diagrams (see both in Fig. 10.1). The action diagram, leveraging weights and ASIs as determined by MUSA, pinpoints priorities for enhancement. The improvement diagram, incorporating AIIs and ADIs, identifies the scope and magnitude of potential improvements.

Unlike traditional applications of MUSA, our study aims to present a holistic overview by integrating action-related metrics of different areas into a single action diagram and improvement diagram. This approach provides a comprehensive, bird's-eye view for recommendations, based on the relative performances across all areas. It is important to note that these recommendations are based on comparative performances, as illustrated in the diagrams where the axes' cutoff levels are recalculated to represent the centroid of all data points.

The overall relative action diagram (Fig. 10.1) organises results into four categories, based on how well different aspects of the cargo bike service are performed (ASIs) and how important these aspects are to citizens (weights):

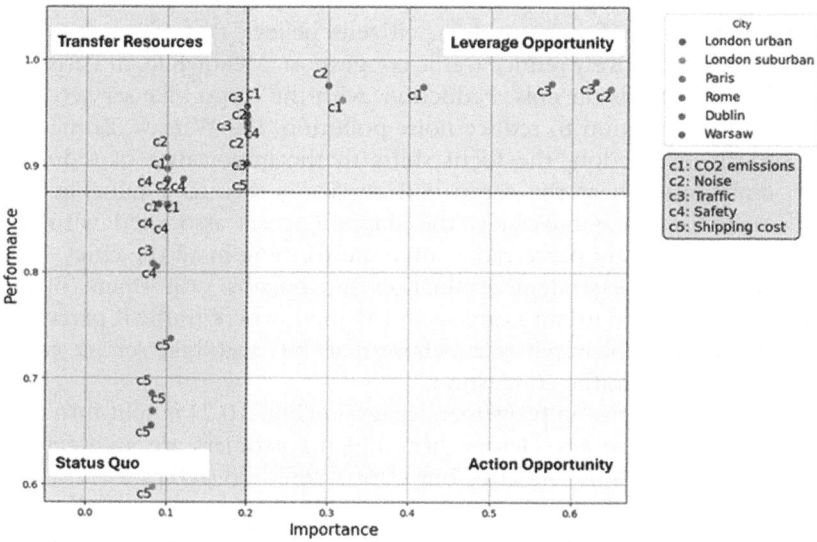

Fig. 10.1 MUSA overall relative action diagram

- *Status Quo:* Identified by low performance and low importance, indicating areas where current performance meets expectations, rendering intervention unnecessary.
- *Leverage Opportunity:* Characterised by high performance and high importance, signalling strengths that could be capitalised on as competitive advantages.
- *Transfer Resources:* Denotes high performance but low importance, suggesting a potential misallocation of resources that could be optimised.
- *Action Opportunity:* Marked by low performance yet high importance, highlighting critical areas needing urgent improvement.

The analysis of citizen feedback from all five cities shows a consistently high level of satisfaction with the cargo bike service. Notably, no criteria are identified as both high importance and low performance, suggesting no immediate need for action. However, this result indicates an opportunity to focus on areas where the cargo bike service has a competitive advantage. For example, in urban London and Dublin, where traffic is perceived as a significant problem, citizens believe that the cargo bike service could effectively reduce traffic congestion. Meanwhile, in Paris and Warsaw, the focus is on noise reduction, with the cargo bike service seen as a beneficial solution to reduce noise pollution. For Warsaw, Rome and the suburb of London, the focus shifts to the importance of reducing CO_2 emissions, where the cargo bike service is seen as a valuable tool of decarbonisation. Conversely, the shipping costs associated with the cargo bike service are perceived as underperforming in all the cities. This suggests a need for strategic evaluation and possible adjustment of the pricing structure. In urban London and Dublin, where traffic is perceived as a significant problem, citizens believe that the cargo bike service could effectively reduce traffic congestion.

The overall relative improvement diagram (Fig. 10.2) is split into four parts, based on two key factors: how much customers are asking for a change (ADI) and how effective our efforts could be (AII):

- *First Priority:* Criteria in this section are characterised by high demand and high effectiveness. This section is where we see values (i.e. sustainability criteria) that citizens most desire and will be most impactful if adopted, yet, aren't too hard to implement.

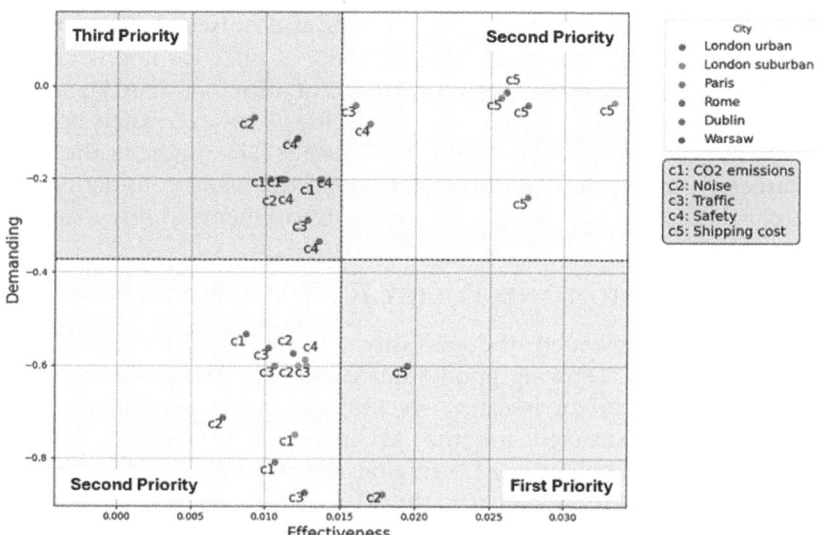

Fig. 10.2 MUSA overall relative improvement diagram

- *Second Priority*: Criteria in this area denote either high demand and low effectiveness, or low demand and high effectiveness. These areas require a balanced approach, which means, the policymakers need to decide carefully where to put resources.
- *Third Priority*: Characterised by low demand and low effectiveness. This last part points out areas that might not be worth the immediate effort because they don't make a huge difference right now and are tough to tackle.

This improvement diagram provides key insights into each city's areas that need improvement. For example, in Warsaw, the shipping cost emerges as a primary concern, highlighting it as a top priority area for improvement. On the contrary, noise reduction is a top priority in Paris. In both cases, these concerns are characterised by a low level of demand from citizens but offer significant room for improvement.

It is also evident that the **delivery cost** for sustainable cargo bike delivery services remains a shared concern for all surveyed citizens across all five cities. In contrast, issues such as **traffic and noise**, while less prioritised by citizens, present challenges in terms of effective improvement. Although preferences vary between cities for different criteria, **safety** stands out as a high-demand criterion in all cities. However, safety appears to be a difficult criterion to improve effectively. This suggests the need for targeted strategies to address safety concerns while considering the inherent difficulties in making substantial improvements in this area.

Conclusion and Policy Recommendations

This study demonstrated the necessity of SSH-STEM collaboration among researchers. Drawing upon SSH researchers' knowledge of cargo bike adoption in urban logistics, an evidence-based hypothetical case scenario was constructed for the MUlticriteria Satisfaction Analysis (MUSA) method application. Leveraging the STEM researchers' expertise in and optimisation statistics, the MUSA method was chosen as a tool to analyse citizens' preferences towards a proposed sustainable solution. This chapter provides new insights to the scholarship on last-mile delivery on the preferences for cargo bike delivery. Our empirical findings suggested that citizens in the selected European capital cities value environmentally sustainable last-mile delivery options. This has echoed the findings from Caspersen and Navrud (2021) that consumers care about the environmental impacts of the last-mile delivery they generate. Moreover, previously mentioned delivery cost for sustainable cargo bike delivery services remains a shared concern for all surveyed citizens. This finding provides a more nuanced understanding of the willingness-to-pay for an alternative last-mile delivery concept (Hagen & Scheel-Kopeinig, 2021) across different cities.

The MUSA-based survey tool offers an alternative approach to gauge public opinions in sustainable logistics policy decision-making. Stakeholder engagement is an integral part of the European Union's sustainable logistics policy framework. The EU promotes public participatory processes and has developed regulatory frameworks, guidelines, and tools to ensure effective stakeholder involvement and increase policy legitimacy (van der Linde et al., 2021). One of the challenges to engaging citizens and consumers in urban logistics policy consultation is to provide sufficient information for the participants in a limited time and space (Tuomala et al., 2023). To overcome this challenge, in this study, to engage citizens from different cities and cultural backgrounds to participate in the same policy evaluation on sustainable cargo bike delivery, we created an evidence-based hypothetical scenario of a cargo bike delivery service to make it easy and accessible for the general population to partake. The MUSA survey procedure is easy and straightforward, facilitating participants in the effortless completion of their responses.

The MUSA-based public perception survey tool proposed in this study offers a methodological guideline for mobility and logistics policymakers (such as the national transport department, local authorities, transport planners), allowing them to map the public's perception and attitude towards sustainable last-mile delivery solutions. Different from the traditional survey descriptive analysis, the MUSA analysis produces the "action diagram and improvement matrix diagram" as the key output of this policy tool (as in Fig. 10.2). The diagram offers a bird's-eye view of the citizens' sustainability prioritisation towards the hypothetical cargo bike delivery service. The matrix diagram can capture areas that require improvement for each city. Moreover, this model considers citizens' perspectives and specific needs in cities of varying scales, making its recommendations both transferable and scalable. The adaptability of this survey tool can be applied across a wide range of scenarios, enhancing the effectiveness of sustainable last-mile delivery solutions.

Different cities have different sustainability priorities when it comes to the sustainable urban freight transport. City managers and logistics practitioner could offer tailored policies and market proposition to address citizens' needs. Reducing carbon emissions as a sustainable goal has been demonstrated by citizens of all cities, but it was not shown as the top priority. For example, by choosing sustainable cargo bike delivery services, citizens in London and Dublin hope to ease the traffic congestion; citizens from Paris and Warsaw hope to reduce the noise. The MUSA Average Satisfaction Indices help to visualise and prioritise the perception and attitudes of citizens towards sustainable last-mile delivery initiatives, thus providing evidence-based support for local authorities and city managers alike to gauge a more nuanced view of their community and neighbourhood.

References

Amling, A., & Daugherty, P. J. (2020). Logistics and distribution innovation in China. *International Journal of Physical Distribution & Logistics Management, 50*(3), 323–332.

Cairns, S., & Sloman, L. (2019). *Potential for e-cargo bikes to reduce congestion and pollution from vans in cities*. Transport for Quality of Life Ltd.

Carracedo, D., & Mostofi, H. (2022). Electric cargo bikes in urban areas: A new mobility option for private transportation. *Transportation Research Interdisciplinary Perspectives, 16*, 100705.

Caspersen, E., & Navrud, S. (2021). The sharing economy and consumer preferences for environmentally sustainable last mile deliveries. *Transportation Research Part D: Transport and Environment, 95*, 102863.

Czepkiewicz, M., Jankowski, P., & Zwoliński, Z. (2018). Geo-questionnaire: A spatially explicit method for eliciting public preferences, behavioural patterns, and local knowledge–an overview. *Quaestiones Geographicae, 37*(3), 177–190.

Dablanc, L. (2019). *E-commerce trends and implications for urban logistics. Urban logistics. Management, policy and innovation in a rapidly changing environment* (pp. 167–195). Kogan-Page.

Dablanc, L. (2023). Urban logistics and COVID-19. In *Transportation amid pandemics* (pp. 131–141). Elsevier.

Dablanc, L., Morganti, E., Arvidsson, N., Woxenius, J., Browne, M., & Saidi, N. (2017). The rise of on-demand 'Instant Deliveries' in European cities. In *Supply Chain Forum: An International Journal, 18*(4), 203–217.

European Commission. (2024). *Sustainable urban mobility—European Commission.* https://transport.ec.europa.eu/transport-themes/urban-transport/sustainable-urban-mobility_en. Accessed 29 January 2024.

Foltýnová, H. B., Vejchodská, E., Rybová, K., & Květoň, V. (2020). Sustainable urban mobility: One definition, different stakeholders' opinions. *Transportation Research Part D: Transport and Environment, 87*, 102465.

Gonzalez-Calderon, C. A., Posada-Henao, J. J., Granada-Muñoz, C. A., Moreno-Palacio, D. P., & Arcila-Mena, G. (2022). Cargo bicycles as an alternative to make sustainable last-mile deliveries in Medellin, Colombia. *Case Studies on Transport Policy, 10*(2), 1172–1187.

Grigoroudis, E., & Siskos, Y. (2002). Preference disaggregation for measuring and analysing customer satisfaction: The MUSA method. *European Journal of Operational Research, 143*, 148–170.

Grigoroudis, E., & Siskos, Y. (2010). *Customer satisfaction evaluation*. Springer.

Hagen, T., & Scheel-Kopeinig, S. (2021). Would customers be willing to use an alternative (chargeable) delivery concept for the last mile? *Research in Transportation Business & Management, 39*, 100626.

Hopkins, D., & McCarthy, A. (2016). Change trends in urban freight delivery: A qualitative inquiry. *Geoforum, 74*, 158–170.

Huang, H., Corrente, S., Kiba-Janiak, M., Siraj, S., & Zhang, X. (2024). Survey data for multicriteria satisfaction analysis of Cargo bike last-mile delivery in European cities. *Zenodo.* https://doi.org/10.5281/zenodo.11401064

Llorca, C., & Moeckel, R. (2021). Assessment of the potential of cargo bikes and electrification for last-mile parcel delivery by means of simulation of urban freight flows. *European Transport Research Review, 13*(1), 33.

Maltese, I., Marcucci, E., Gatta, V., Sciullo, A., & Rye, T. (2023). Challenges for public participation in sustainable urban logistics planning: The experience of Rome. In *Public participation in transport in times of change* (pp. 77–95). Emerald Publishing Limited.

Rodrigue, J. P., Dablanc, L., & Giuliano, G. (2017). The freight landscape: Convergence and divergence in urban freight distribution. *Journal of Transport and Land Use, 10*(1), 557–572.

Rose, W. J., Bell, J. E., Autry, C. W., & Cherry, C. R. (2017). Urban logistics: Establishing key concepts and building a conceptual framework for future research. *Transportation Journal, 56*(4), 357–394.

Ryghaug, M., Subotički, I., Smeds, E., von Wirth, T., Scherrer, A., Foulds, C., Robinson, R., Bertolini, L., Beyazit İnce, E., Brand, R., Cohen-Blankshtain, G., Dijk, M., Pedersen, M. F., Gössling, S., Guzik, R., Kivimaa, P., Klöckner, C., Nikolova, H. L., Lis, A., … Wentland, A. (2023). A Social Sciences and Humanities research agenda for transport and mobility in Europe: Key themes and 100 research questions. *Transport Reviews, 43*(4), 755–779.

Tsavachidis, M., & Le Petit, Y. (2022). Re-shaping urban mobility–Key to Europe's green transition. *Journal of Urban Mobility, 2*, 100014.

Tuomala, V., Aminoff, A., & Gammelgaard, B. (2023). *Consumer and citizen perspectives on sustainability in last-mile deliveries.* Abstract from The 35th NOFOMA Conference 2023, Helsinki – Espoo, Finland.

van der Linde, L. B. A., Witte, P. A., & Spit, T. J. M. (2021). Quiet acceptance vs. the 'polder model': Stakeholder involvement in strategic urban mobility plans. *European Planning Studies, 29*(3), 425–445.

Open Access This chapter is licensed under the terms of the Creative Commons Attribution 4.0 International License (http://creativecommons.org/licenses/by/4.0/), which permits use, sharing, adaptation, distribution and reproduction in any medium or format, as long as you give appropriate credit to the original author(s) and the source, provide a link to the Creative Commons license and indicate if changes were made.

The images or other third party material in this chapter are included in the chapter's Creative Commons license, unless indicated otherwise in a credit line to the material. If material is not included in the chapter's Creative Commons license and your intended use is not permitted by statutory regulation or exceeds the permitted use, you will need to obtain permission directly from the copyright holder.

PART V

Conclusion

CHAPTER 11

Recommendations for Future Interdisciplinary Collaborations Within Transport and Mobility

Marianne Ryghaug, Tomas Moe Skjølsvold, Imre Keseru, and Samyajit Basu

Abstract This book explores the potential of interdisciplinary collaborations between Social Sciences and Humanities (SSH) and Science, Technology, Engineering, and Mathematics (STEM) to address challenges in transport and mobility. Through a series of experimental research teams,

M. Ryghaug · T. M. Skjølsvold
Department of Interdisciplinary Studies of Culture, Norwegian University of Science and Technology, Trondheim, Norway
e-mail: marianne.ryghaug@ntnu.no

T. M. Skjølsvold
e-mail: tomas.skjolsvold@ntnu.no

I. Keseru (✉) · S. Basu
House of Sustainable Transitions (HOST), Mobilise Mobility and Logistics Research Group, Vrije Universiteit Brussel, Brussels, Belgium
e-mail: imre.keseru@vub.be

S. Basu
e-mail: samyajit.basu@vub.be

© The Author(s) 2024
I. Keseru et al. (eds.), *Strengthening European Mobility Policy*,
https://doi.org/10.1007/978-3-031-67936-0_11

the authors aimed to integrate diverse disciplinary insights to produce innovative policy recommendations. Despite encountering difficulties and traditional disciplinary boundaries, the project highlights the importance of institutional support, adequate resources, and long-term commitment to fostering effective interdisciplinary research. The concluding chapter reflects on these experiences and offers recommendations for enhancing future interdisciplinary efforts in this field.

Keywords Interdisciplinary collaboration · Financial incentives · Institutional requirements · Radical interdisciplinarity

Introduction

This book sets out from a normative and experimental starting point: We assumed that if scholars from across the divide of SSH and STEM work together to address challenges within transport and mobility, we would likely end up with a combination of better and surprising insights combined with policy recommendations that in new and holistic ways challenge the modus operandi of transport and policymaking in the EU. To unlock this potential, we established a series of interdisciplinary research teams across the continent, who were tasked precisely with combining SSH and STEM to analyse a phenomenon and come up with policy recommendations. Against this backdrop there are many ways to read the preceding chapters. On the one hand, it is possible to read this book as a failure: many of the book chapters come across as relatively traditional analyses, where it may be difficult to see the explicit inter-disciplinary contribution. Another, and more constructive, reading is to view the book as a process-oriented experiment, valuing the process and learning from it, as well as the seeds that the process may have yielded in a long-term perspective. In this concluding chapter, we will discuss the sum of what this experimental book can teach us about SSH-STEM collaborations, about the transport and mobility field, as well as the challenges of translating knowledge into policy recommendations.

On Interdisciplinary Collaboration

In this book, we have tried to push a multi- and interdisciplinary research agenda. This means that we break with what has been the dominant organisation of knowledge and science in the twentieth century, namely a traditional form of disciplinarity (Klein, 2010). Multidisciplinary approaches aim to bring disciplines together, but typically end up with disciplines working side-by-side, rather than in an integrated and collaborative way. Interdisciplinarity, on the other hand, is based on not only juxtaposing knowledge from different disciplines, but concerns the integration of data, methods, tools, concepts, theories, and/or perspectives from multiple disciplines in order to answer a question, solve a problem or address a topic that is too broad or complex to be dealt with by one discipline (Klein, 2010). Arguably, transportation and mobility are examples of such complex topics. Interdisciplinarity has for the last decade expanded into a heterogeneity of practices and forms. Today it ranges from borrowing tools and methods across academic fields, to forming new fields and inter-disciplines. So, what form of interdisciplinary do we find represented on the pages of this edited volume?

In the chapters of this book, we identify several interesting approaches to interdisciplinary work where the authors strive to bridge disciplines and conduct disciplinary work across SSH and STEM.

Despite many good attempts, we still observe that disciplines continue to speak with separate voices and retain original identity. This should not be a surprise. Previous research on the practices of interdisciplinarity illustrates that interdisciplinary is a challenging endeavour. While our book represents a small-scale, one-off effort to nurture new practices, to make new and long-lasting interdisciplinary collaborations across SSH-STEM requires consideration of institutional context, availability of resources, and time.

For instance, the common assumption that complex problems must be solved by integrating interdisciplinary solutions have often proven to "melt under closer inspection" (Winskel, 2014, p. 78), as problems could also be tackled by partial or specialised knowledge or solutions that are integrated or collated afterwards. In this book, authors were asked to design policy recommendations based on their interdisciplinary knowledge process. This then mimics typical policymaking processes, as policy often demands some kind of integrated approach and do not settle with partial or specialised answers. This would lead policymakers themselves to have to do the integration work. Designing policy recommendations that foster integration or a holistic approach, however, is a very ambitious task. With this as a backdrop, our project can be critiqued as somewhat naïve.

What Do the Chapters Say About Interdisciplinary Collaborations?

The character of the interdisciplinary collaborations is described to a varying degree in different chapters. There are examples of discussions that highlight the merit of integrating insights, e.g. as Chapter 6 (Lieszkovszky et al.) attempts to bring transport engineering and social science disciplines together in a discussion about demand responsive transport solutions. In Chapter 5 (Krumnikl et al.), the technological solution at hand is taken for granted in an exploration of barriers to implementation—STEM researchers investigated the efficiency of new electric buses, while SSH scholars provided broader contextual research and focussed on public perception. Chapter 9 (Michailidis et al.), arguably does things the other way around. Here, the problem addressed is how to design policy bottom-up through co-creation, while the case at hand is artificial intelligence. These three examples, then, demonstrate different approaches to interdisciplinarity in our book.

Reflections on SSH-STEM Collaborations

This book project has been an experiment in facilitating interdisciplinary research collaboration. One way that authors have solved the challenge of doing interdisciplinary research is by using SSH-related tools and methods to answer a STEM problem formulation, and vice versa and methods from other disciplines to answer to the problem formulation, as illustrated in the previous paragraph. This type of interdisciplinarity, where social science plays a subservient or "gap-filling" role, is rather common, but contested by some as it might re-enforce stale understandings of what the underlying problems are, rather than opening up for more radical explorations (Winskel, 2018: 78).

Nevertheless, the research teams had limited time to carry out the research. They collected and analysed data and wrote the first draft in a course of 6 months. The collaborations assumed that the members of the team had not worked together before, which required time and effort to develop a common understanding of ways of collaboration and an understanding of different languages that the SSH and STEM communities may speak. While good collaboration, especially across different disciplines, may take years to build, our authors endeavoured to a fast track to interdisciplinarity.

In doing so, the book project experiment has probably been successful in making transport and mobility scholars look beyond their normal or traditional disciplinary way of defining research questions and problems, though we have probably not revolutionised the transport and mobility field. On the other hand, we should not be too cynical as it is difficult to predict the effects that the experiment may have in the longer run. The book project experiment and the activities of the chapter teams may have longer-term impact in broadening the horizons of those participating. Perhaps the fruits of this experiment may be seen in future transport and mobility projects and future research collaborations. For this to happen, it surely needs to be matured and nurtured as we suggest in the policy recommendations below.

Challenges and Opportunities for Interdisciplinarity

Interdisciplinarity is not easy. In this book, scholars with different levels of experience and training in the disciplines are probed to work across SSH and STEM, thus, performing a relatively rare form of 'radical interdisciplinarity'. To come up with policy recommendations based on such an approach is difficult, as it demands some form of *integration* of the scientific output. Policy recommendations also demand converting knowledge into action-propositions. Even more basic and 'cognate' interdisciplinarity is hard to conduct and in need of specific nurturing, in terms of institutional embedding, time, and resources. In deep academic structures, such work still has a hard time. Universities are built around traditional disciplines, and funding bodies massively over-fund STEM compared to SSH (Silvast & Foulds, 2022).

Another challenge related to formulating policy recommendations at the EU level, is that local and regional transport is regulated at the local levels with just general EU policies and guidelines in place (such as the guidelines for preparing sustainable urban mobility plans), while local authorities may choose their own approaches to organising public transport. Most of the chapters in this book report on research that was conducted in relation to such local or city-level policies. These may, in practice, prove difficult to connect to the EU level, which remains more abstract and difficult to grasp for the researchers.

It is both time-consuming, difficult and honestly—often frustrating—to collaborate across disciplines. At the same time, successful interdisciplinary groups, projects, or networks are vulnerable within contemporary knowledge production institutions. It often comes across as a no-brainer that one should collaborate across disciplines, but in practice the epistemic, ontological, and practical challenges tend to push researchers back to their well-established disciplines. Thus, if one wants to have more interdisciplinary collaboration between SSH and STEM, this needs specific focus and attention in calls, longer-term financing, and the training of scholars. Therefore, our concluding policy recommendations are:

- Link the strategic goals of interdisciplinarity to financial incentives and methods for governing.
- Provide enough time and space (and coordinating functions) to allow for interaction and mediation between researchers.
- Consider consortia sizes, as smaller and more tightly interwoven research teams are more successful for inter- and transdisciplinary research compared to large and more loosely organised teams or networks.

References

Klein, J. T. (2010). A taxonomy of interdisciplinarity. In R. Frodeman, J. T. Klein, & C. Mitcham (Eds.), *The Oxford handbook of interdisciplinarity* (pp. 15–30). Oxford University Press.

Silvast, A., & Foulds, C. (2022). *Sociology of interdisciplinarity: The dynamics of energy research*. Springer Nature. https://doi.org/10.1007/978-3-030-88455-0

Winskel, M. (2018). The pursuit of interdisciplinary whole systems energy research: Insights from the UK Energy Research Centre. *Energy Research & Social Science, 37*, 74–84.

Open Access This chapter is licensed under the terms of the Creative Commons Attribution 4.0 International License (http://creativecommons.org/licenses/by/4.0/), which permits use, sharing, adaptation, distribution and reproduction in any medium or format, as long as you give appropriate credit to the original author(s) and the source, provide a link to the Creative Commons license and indicate if changes were made.

The images or other third party material in this chapter are included in the chapter's Creative Commons license, unless indicated otherwise in a credit line to the material. If material is not included in the chapter's Creative Commons license and your intended use is not permitted by statutory regulation or exceeds the permitted use, you will need to obtain permission directly from the copyright holder.

Afterword 1: From Many Hands Problem to Unconscious Assumptions: Transforming Our Governance Systems

Miloš N. Mladenović

This interdisciplinary book provides findings on several important topics. On the one hand, looking outwards at our places of (im)mobilities, it suggests changes in our transport technologies and improved integration in the mobility system. On the other hand, looking within our governance systems, the book suggests improvements in methods (e.g., for design or evaluation) or data collection (e.g., on school-related travel). Furthermore, the book also provides suggestions for policy integration across sectors and the development of inclusive policy processes, while underlining cooperation across different actors. Thus, this book provides insights for both research and practice by emphasising that changes in our shared places are contingent upon changes in our collective governance systems.

As highlighted in several book chapters (e.g. Chapter 3 [Lait et al.,], Chapter 9 [Michalidis et al.], and Chapter 10 [Huang et al.]), the problem in front of us is one of many responsible hands for transformation. We need to have those many hands coordinating their actions grounded in evidence and vision. As already mentioned in the foreword to this book,

M. N. Mladenović
Department of Built Environment, Aalto University, Espoo, Finland
e-mail: milos.mladenovic@aalto.fi

this is easier said than done. Here, in addition to the many hands problem (Thompson, 1980), it is good to recall an ancient Indian parable "The Blind Man and the Elephant". The poem tells the story of several blind men who have a first encounter with an elephant. They attempt to understand its nature by feeling out different parts of its body. Each man perceives the elephant differently based on the specific part they touch (e.g., the trunk, the tail, the tusk), leading to conflicting interpretations of what the elephant is like. The crux of the story is that we face limitations when we try to collectively understand complex phenomena. Now imagine that those blind men are also supposed to steer the elephant. This combined many hands and many eyes problem is exactly what happens in our currently siloed and multi-layered governance systems.

So, the obvious next thought is—well, we need to improve knowledge and action integration, as highlighted in Chapter 3 by Lait et al. in terms of education and transport policy. Such integration obviously needs a range of changes, beyond methods, data, or process design. Starting from available resources and legislation, through organisational structures to power redistribution, the governance system change quickly becomes a political and even a moral question. This certainly does not make change easier. Fortunately, we already have a plethora of policy and governance theories to enlighten the way—from bounded rationality to punctuated equilibrium theory, multiple streams, and policy learning, to name the few (Cairney, 2020).

Going beyond traditional lessons from transport or policy studies, I would propose two additional key considerations. First, take a moment to think about how much research we have about humans in our mobility systems. We know very well that travelling has to do with human attitudes, habits, biases, experience, and norms—among other things. So, what is stopping us from seeing also the governance system filled with humans to also comprise of such aspects? For that change in perspective, we can draw lessons from organisational studies to introduce underlying assumptions and meanings into our systemic understanding (Olin & Mladenović, 2024). The challenge here is to understand that many of these collective assumptions are rather unconscious, developed over long spans of time in both the culture of a society at large and within a specific organisational culture. Such a perspective would certainly help us to further understand and act upon the intervention points for governance system change.

The second consideration requires us to rethink our dominant narratives (Te Brömmelstroet et al., 2022). In particular, we need to move away from the current Eurocentric, colonial, exploitative, negatively discriminatory, and growth-obsessed paradigm. Such narrative is at the core of our current pathway into the multiple crises of human and planetary well-being, with irreversible damage for the life of this bounded planet. To be able to rethink this paradigm, we first and foremost need courage across our governance networks. However, this courage needs to be balanced with mutual empathy, which is even more important in times of societal transformation. Ultimately, I would like to turn the mirror of this reflection towards my colleagues in academia. Despite academia often being the anchor of stability through knowledge preservation, we have to actively question our own privilege and power (Ryghaug et al., 2023). Part of that questioning involves also deconstructing traditional divisions and lack of collaboration between SSH and STEM fields. Only then, we will be able to shift to a new level of collective meanings, and to have actual hope for a just societal transformation.

References

Cairney, P. (2020). *Understanding public policy: Theories and issues* (2nd ed.). Bloomsbury Publishing.

Olin, J. J., & Mladenović, M. N. (2024). Unpacking the cultural aspects of transport automation governance in Finland: An interview study. *Journal of Transport Geography, 117*, 103874.

Ryghaug, M., Subotički, I., Smeds, E., von Wirth, T., Scherrer, A., Foulds, C., Robinson, R., Bertolini, L., Beyazit İnce, E., Brand, R., Cohen-Blankshtain, G., Dijk, M., Pedersen, M. F., Gössling, S., Guzik, R., Kivimaa, P., Klöckner, C., Nikolova, H. L., Lis, A., Marquet, O., ... Wentland, A. (2023). A Social Sciences and Humanities research agenda for transport and mobility in Europe: key themes and 100 research questions. *Transport Reviews, 43*(4), 755–779.

Te Brömmelstroet, M., Mladenović, M. N., Nikolaeva, A., Gaziulusoy, İ., Ferreira, A., Schmidt-Thomé, K., Ritvos, R., Sousa, S., & Bergsma, B. (2022). Identifying, nurturing and empowering alternative mobility narratives. *Journal of Urban Mobility, 2*, 100031.

Thompson, D. F. (1980). Moral responsibility of public officials: The problem of many hands. *American Political Science Review, 74*(4), 905–916.

Afterword 2: What (About) Now? Complexities, Omissions, and Taking Transitions Seriously

Debbie Hopkins

How to describe the world in 2024? Many scholars, practitioners, policy-makers, and commentators have used terms such as polycrisis, compound crises, or cascading crises to make sense of the intersecting and relational social, economic, political, and ecological challenges facing humans and non-human communities alike, across the world. Maybe what is being viewed as a "crisis" today has always been; perhaps it is only now that we are becoming attuned en-masse to the dire and sustained injustices and inequities that are both cause and consequence of the status quo.

Growing recognition of the powerful role that ways of knowing emanating from disciplines of the social sciences and humanities (SSH) may also play a part. Even at the turn of the century, climate breakdown was treated as a problem to be solved by science. The dominant problem framing all but precluded social scientists (let alone humanities scholars) from sitting at the table. The same could be said for a variety of other ecological challenges. Even the very development of sustainability as a concept, object of study, and policy orientation, happened in such a way that only certain types of knowledge (and thereby discipline) could contribute.

D. Hopkins
Department for Continuing Education, University of Oxford, Oxford, UK

Fast forward to the early 2020s, and huge strides have been taken to challenge dominant framings and show their limitations, to reposition problems as multifaceted, complex, and relational, and to widen participation in terms of (inter-) discipline, region, gender, and more. Yet problems remain. As we speak, the UK government is reducing funding for humanities disciplines across UK higher education institutions—questioning the feasibility of running these courses at many institutions, particularly those serving first-generation and lower-income students. And while the value—and necessity—of interdisciplinary collaborations for global challenges is more widely recognised, many funders remain reluctant to support many of these ambitious projects, often tied into discipline specific units of evaluation.

A further challenge is that of making the knowledge produced by these collaborations usable (should that be the aim) for policy and practice audiences. Within this context, the SSH CENTRE is novel in both its scale and scope; taking on the established ways of thinking and doing research across climate, energy and mobility, and working with—rather than against—STEM. Across the chapters within this book a number of my own priorities emerge, two of which I discuss below. These—for me—represent important themes for progressing a SSH agenda across climate, energy, and mobility.

The first priority is to *move beyond the "usual suspects"*. Thinking on transport transitions to carbon neutrality have largely centred on urban environments (predominantly medium-large cities; e.g., Chapter 10, Huang et al.), and passenger mobility (Chapter 5, Krumnikl et al.). The former is due perhaps to the relative ease and abundance of measures that become available with population density, and the incompatibility of rural areas with dominant transition priorities (e.g., electric mobility; Chapter 4, Lis-Plesińska et al.). The latter reflects not only where most emissions lie, but also where there is an agency for both public and private sector actors, as well as communities, to act (while freight transport decarbonisation, in the UK at least, has by and large been left to private sector interventions).

Yet rural areas (Chapter 6, Lieszkovszky et al.) and road freight (Chapter 10, Huang et al.) are lagging behind in the transition, and risk becoming a hindrance to systemic change. Two immediate issues arise: 1. Neglect of the spatial dimensions of transitions, which defers to and prioritises urban conditions and thereby overlooks distinctive and diverse rural characteristics in which change will happen, and 2. Only understanding freight mobilities through the paradigm of passengers without recognising the workings of logistics and supply chain capitalism.

The second priority is to focus on the *acceptability of sustainable mobility interventions*. Living in Oxford (UK), I am reminded daily of the ongoing and highly contentious—even antagonistic—debate around sustainable mobility interventions, namely low traffic neighbourhoods, zero emission zones, and 15-minute city infrastructure. The introduction of policy to reduce car use and increase active mobility for health and environmental benefits, has become a key fault line in the local community and, interestingly, beyond. But these debates cannot be understood through rational evaluation processes (see Chapter 2 by Alonso et al.), they signal the ways that sustainable mobility policy is experienced by resident groups (just as Lis-Plesińska et al. explore in Chapter 4). In other words, these interventions (such as those introduced in Chapter 5 by e.g., Krumnikl et al.) are political and spatially determined.

There is a pressing need for interdisciplinary engagement as we transition to a post-carbon world. SSH remain critical if we are to overturn the logics and practices that created the conditions for fossil-fuel dependent economies. But what these SSH engagements will look like is yet to be defined.

Afterword 3: Cities in Transition

Lucian Zagan

Cities are increasingly recognised as critical leverage points in combating global warming and climate change. City administrations have competencies in key sectors essential for sustainability transitions, such as waste management, land-use planning, energy in buildings, transport. Being closer to citizens than national governments, city administrations can make more effective decisions and play a vital role in a just transition. The EU Mission on Climate-Neutral and Smart Cities acknowledges the pivotal role of cities in the European ecosystem and aims to create transition pathways for long-lasting impacts by leveraging their unique position. Many cities have developed climate action plans and urban climate governance structures, often implementing more progressive policies than national governments. Cities are focal points for policy and societal action, experimenting with innovative policy and planning methods. They also join national and transnational networks for knowledge sharing and collaborative problem-solving. Innovations from local levels often inspire actions in other cities and influence national and EU policies.

Transport is a key sector in the transition to climate neutrality, being the second-largest source of greenhouse gas emissions in Europe. As part of their climate policy goals and facing the negative effects of excessive car use in the most direct way, cities have turned towards sustainable urban

L. Zagan
Eurocities, Brussels, Belgium

mobility. Sustainable Urban Mobility Plans (SUMP) were proposed as a tool to drive the modal shift towards public transport, cycling, walking, and shared services, and they were widely adopted by cities across Europe as an integrative and efficient form of mobility planning and driving the mobility transition. The European Commission widely promoted the SUMP concept starting with its first definition in the 2013 Urban Mobility Package, funds were made available for urban mobility projects, and a large knowledge base was created through relevant research and innovation. At the same time, a variety of local actions were developed, tested, and scaled up, the mobility landscape changing significantly in many cities across Europe. Still, systemic change has not yet been achieved and transitions have been slow and geographically uneven. Indeed, a special report from the European Court of Auditors (2020: 37) states that "there is no clear indication that cities are fundamentally changing their approaches" and "[t]here is no clear trend towards more sustainable modes of transport."

One of the key challenges lies with the very nature of transport. Transportation is ultimately a derived demand that stems from the organisation of city life and how the city responds to the needs of its residents. Trips are generated from decisions and the way things are organised in sectors other than transport, e.g. education, health, retail, or tourism. A successful mobility transition to advance towards climate neutrality requires establishing cross-sector collaboration to avoid unnecessary trips and shift towards sustainable modes for the remaining transportation needs (see Chapter 3, Lait et al.). Cross-sectoral collaborative transformative strategies and policy implementation need to be established as a way of driving the mobility transition and climate mitigation. Evidence-based and critically informed reflection are essential to meet visions with action and to ensure policy coherence towards specific goals such as sustainable mobility or climate mitigation. Quite often in sustainability discussions, lacking clear evidence or ignoring it, we see ideology taking over and driving relevant discussions or policies reversed or marred by—sometimes subtle, sometimes straightforward—incoherence. Improving monitoring and evaluation (see Chapter 2, Alonso et al.), diversifying methods and tools, including automated, artificial intelligence-driven ones (see Chapter 8, Tori et al. and Chapter 9, Michailidis et al.), testing policy incentives and other relevant behavioural tools (see Chapter 7, Petrakis et al.), identifying new data sources made possible by technology are all very relevant in guiding policymaking and backing up decision-making with evidence and analysis.

Systemic transitions are, by definition, work in progress. Interdisciplinary research and innovation, partnerships with academia, engaging in knowledge sharing, capacity building, and creating new roles are vital for city administrations to transition to sustainable mobility and climate neutrality. Testing and piloting new collaborative approaches through research and innovation allow for gathering evidence, making refinements, and iterating before scaling up (see Chapter 11, Ryghaug et al.). Creating space for experimenting helps dealing with uncertainty or lack of consensus on effective solutions. Living labs for ongoing experimentation are essential to build an evidence base and develop localised transition pathways, recognising the significance of context. Additionally, research and innovation are crucial to understanding the success or failure factors of past trends, societal changes, and governance capacities in urban transitions. Driving factors are complex, and technological solutions alone will not achieve a low-carbon future. Transformative policies need fresh perspectives and blending insights from SSH and STEM research, as illustrated by the research presented in this book.

References

European Commission. (2013). *Communication from the Commission to the European Parliament, the Council, the European Economic and Social Committee and the Committee of the Regions: Together towards competitive and resource-efficient urban mobility (COM/2013/0913 final)*. https://eur-lex.europa.eu/legal-content/EN/TXT/?uri=CELEX:52013DC0913

European Court of Auditors. (2020). *Sustainable Urban Mobility in the EU: No substantial improvement is possible without Member States' commitment (Special report 06/2020)*. https://www.eca.europa.eu/Lists/ECADocuments/SR20_06/SR_Sustainable_Urban_Mobility_EN.pdf

Afterword 4: Sustainable Mobility and Systemic Change: The Power of Collaborative Governance

Oliver Greenfield and Roberta Dall'Olio

The author teams in this collection offer pieces of the new to-be-assembled mobility puzzle, which we have read with interest. In this Afterword, we would like to offer some reflections on puzzle-making—not necessarily about any of the individual pieces (technologies, infrastructures, policy measures)—but what you need as a minimum for them to come together. Put differently, we want to reflect on the conditions for change.

From our experience and observations as directors of a green economy coalition and a sustainable development association, respectively, we start with what might seem like a truism: the governance system of today is not equipped for the mobility system of tomorrow. It was purpose-built for a different world. This basic observation might be simple. In fact, most of the authors seem to take this observation as a given. However, its implications, in each region and each case, run wide and deep. Therefore, we would like to dwell on it to help conclude this book.

O. Greenfield
Green Economy Coalition, London, UK
e-mail: oliver.greenfield@greeneconomycoalition.org

R. Dall'Olio
European Association of Development Agencies, Brussels, Belgium
e-mail: roberta.dallolio@eurada.org

For any mobility challenge or opportunity, before businesses, governments or citizens can act, they need to understand the context in which they would like to make a change. Fortunately, this collection includes various chapters that offer (new) ways of how to inform oneself, whether that be through AI (Chapter 9, Michailidis et al.) or innovative surveying (Chapter 10, Huang et al.). We want to draw special attention to the contribution of Alonso et al. (Chapter 2) on the evaluation of public policies. It directly speaks to the fact that knowledge is distributed across institutions and stakeholders, which have parcelled up reality—and thus their knowledge about it—in ways that do not necessarily line up with the changing world. This can hinder progress. In fact, the societal and business cases for change are hard to define and attribute, because the returns and benefits are typically spread across industries and government departments. For example, bikes offer health and air quality improvements and a reduction in CO_2 but no road tax revenue. So, should cycle lanes be funded by health departments?

This example highlights how knowledge-building needs to integrate multiple perspectives. Any investment into a new organisation of mobility is going to have impact on other domains too. To assess these impacts, policymakers need to switch lanes and build an integrated, holistic approach. The consequence, as we see it, is that informing oneself need not be a one-way street. One party doing all the "asking" would limit the range of perspectives and therefore risk not meeting the challenge of new systemic needs. Sourcing and distributing intelligence among all stakeholders gives one a better chance of getting a more complete overview of the situation. In fact, and more profoundly, data gathering itself cannot be separated from how the problem is defined, which is ideally itself a group effort.

We have some examples of how that looks like in practice. The first is to draw up a pact. A strong example is the Pact for Work and Climate from the Emilia Romana region in Italy. This collaboration for sustainable development of the region involved quadruple helix partners: research institutions, trade unions, local administrations, and more. They agreed on 5-year objectives, an operational plan, and on the indicators to monitor the results. The Pact thus builds a process in which collaboration can take place. Each participating organisations has different tools in their toolbox that they can contribute to the overall plan.

A useful tool in such collaborations is the "Definitions of Success", which the Green Economy Coalition have used in their studies. This tool leads to a map of what success looks like for whom. Thus, successful changes to mobility might mean improved access to places and things for citizens. For businesses, it may mean access to new markets, opportunities for R&D, and continued viability and profitability. Governments might profit from creating a new tax base. There are also generalised benefits, like significant reduction in CO_2, particulates, noise and waste, reduction in material use. This is investigated for example in Chapter 6 (Lieszkovszky et al.) in the context of demand responsive transport where many private operators try to work with governments to make services financially and operationally feasible with less or more success.

We have seen that getting input from a range of stakeholders is more important for another reason: you build a network that you can activate during implementation. The key, during the initial stages of a project or trajectory, is to not only ask stakeholders what they want, but also what they can do, to reach that objective. That can help build a process that can sustain effective collaboration. These points are discussed in Chapter 9 (Michailidis et al.) and Chapter 10 (Huang et al.) of the book from innovative perspectives.

To sum up: systemic change requires a systemic—or holistic—perspective; a holistic perspective requires input from multiple stakeholders; and organising input from multiple stakeholders can be the basis of an implementation plan with a shared vision and shared responsibility.

Index

A
Accessibility, 5, 8, 18, 46, 52, 119, 121, 122
Artificial intelligence (AI), 8, 17, 75, 89, 105, 116, 119, 124, 150, 164

B
Benchmarking, 7, 20, 21
Best practice, 21

C
Cargo bikes, 8, 132, 133, 135–138, 140, 141
Citizen participation, 119
Climate change, vi, ix, x, 14, 43, 88, 95, 163
CNG buses, 7, 61, 63, 65, 67
Communication campaigns, 19
Cooperative Intelligent Transportation Systems (C-ITS), 18
Cost-benefit analysis (CBA), 7, 19, 88
Cross-sectoral impacts, 30, 32

D
Demand responsive transport (DRT), 7, 74, 76, 169
Door-to-door transport, 77, 81
Driving behavior, 89, 90, 93

E
Ecological impact, 49
Ecologisation, 60
Economic impact, 44, 60
Education policy, 29
Electric buses, 47, 60, 61, 63, 65, 67, 150
Energy efficiency, 18, 28, 89
Equity, 5, 15, 18
Evaluation, 7, 15, 16, 18–22, 96, 119, 121, 123, 138, 141, 155, 160, 161, 168

F
Financial incentives, 7, 89–92, 95, 96, 152

G
Geopolitical risks, 42, 49, 50

I
Inclusion, 29, 32, 118, 120
Integration, vi, 6, 66, 149, 150, 155, 156

K
Key performance indicators (KPI), 7, 16, 21

L
Large language models (LLMs), 105

M
Manufacturing, 42, 47, 48, 50
Mobility planning, 8, 164
Mobility practices, 42
Modelling, 29, 105
Multilevel governance, 60, 67

P
Participatory evaluation, 123
Participatory planning, 8
Policy assessment, 16
Policy conflict, 30, 35
Policy coordination, 28, 34–37
Policy interventions, 109
Public opinions, 140
Public transport, 14, 18, 19, 31, 46, 49, 60–63, 65, 66, 68, 74–79, 81, 109, 116, 124, 164

R
Regional strategies, 60, 61, 67

Ride-sourcing, 82
Road safety, 18, 88, 89, 91, 93, 95
Route optimisation, 68, 79
Rural mobility, 75

S
Safe pass, 90, 92, 94, 96
Sensor data, 17
Social acceptance, 18
Social cost-benefit analysis, 7, 89
Socio-economic feasibility, 90, 91, 95
Stakeholder engagement, 131, 140
Subsidies, 48, 60, 66, 68
Supply chains, 42, 44, 45, 47, 50, 131, 161
Sustainable urban mobility, 7, 62, 67, 164
Sustainable urban mobility plan (SUMP), 62, 75, 104, 131, 151, 164

T
Telematics, 89–92, 95, 96
Tourism, 16, 79, 164
Traffic simulation, 17
Transition, v, vi, xi, 4–7, 30, 36, 42, 43, 45, 48, 50, 51, 60, 67, 68, 75, 160, 163–165
Twitter, 105, 106

U
Urban logistics, 130–132, 140

V
Vehicle insurance policies, 7, 90, 95, 96
Vision zero, 96
Vulnerable people, 18, 79

SPRINGER NATURE

GPSR Compliance

The European Union's (EU) General Product Safety Regulation (GPSR) is a set of rules that requires consumer products to be safe and our obligations to ensure this.

If you have any concerns about our products, you can contact us on ProductSafety@springernature.com

In case Publisher is established outside the EU, the EU authorized representative is:

Springer Nature Customer Service Center GmbH
Europaplatz 3
69115 Heidelberg, Germany

The manufacturer's authorised representative in the EU is Springer Nature Customer Service Centre GmbH, Europaplatz 3, 69115 Heidelberg, Germany. If you have any concerns regarding our products, please contact ProductSafety@springernature.com

Printed and bound by CPI Group (UK) Ltd, Croydon, CR0 4YY

23/03/2026

02076447-0003